Mental Models
Design of User Interaction and Interfaces for Domestic Energy Systems

Mental Models
Design of User Interaction and Interfaces for Domestic Energy Systems

Kirsten M.A. Revell and Neville A. Stanton

CRC Press
Taylor & Francis Group
Boca Raton London New York

CRC Press is an imprint of the
Taylor & Francis Group, an **informa** business

CRC Press
Taylor & Francis Group
6000 Broken Sound Parkway NW, Suite 300
Boca Raton, FL 33487-2742

© 2017 by Taylor & Francis Group, LLC
CRC Press is an imprint of Taylor & Francis Group, an Informa business

No claim to original U.S. Government works

Printed on acid-free paper

International Standard Book Number-13: 978-1-4822-4944-6 (Paperback)
International Standard Book Number-13: 978-1-138-08669-2 (Hardback)

Visit the Taylor & Francis Web site at
http://www.taylorandfrancis.com

and the CRC Press Web site at
http://www.crcpress.com

Printed and bound in the United States of America by
Edwards Brothers Malloy on sustainably sourced paper

Contents

List of Figures... xv
List of Tables... xxi
Preface... xxiii
Acknowledgements... xxv
Authors.. xxvii
List of Abbreviations... xxix

Chapter 1   Introduction 1

    1.1   Who Is This Book For? 1
    1.2   How Different Readers Should Approach Reading the Book?..... 1
    1.3   Background... 1
    1.4   Aims and Objectives/Purpose.. 2
           1.4.1   Overall Hypotheses 2
           1.4.2   Sub-Hypotheses... 3
    1.5   Contribution to Research.. 4
    1.6   Structure of the Book.. 4

Chapter 2   Models of Models: Filtering and Bias Rings in Depiction of
              Knowledge Structures and Their Implications for Design 9

    2.1   Introduction 9
           2.1.1   The Concept of Mental Models as Inferred
                    Knowledge in Cognitive Processing........................... 10
                  2.1.1.1   Johnson-Laird (1983)................................... 11
                  2.1.1.2   Bainbridge (1992).. 12
                  2.1.1.3   Moray (1990).. 12
                    2.1.1.4   Summary of Comparison of Theories
                          of Cognitive Processing 13
    2.2   Importance of Accuracy in Mental Model Descriptions:
           The Development of an Adaptable Framework...................... 15
           2.2.1   Bias and Filtering When Constructing or
                    Accessing Mental Models 15
           2.2.2   Accuracy of Mental Model Content:
                    A Case Study of Kempton (1986) Illustrating
                    the Impact of Methodology 19
                  2.2.2.1   Bias When Accessing Another Person's
                          Mental Model.. 20

2.2.3   Accuracy in Definition: The Perspective from
        Which Data Is Gathered 23
        2.2.3.1   Norman (1983) 24
        2.2.3.2   Wilson and Rutherford (1989) 26
        2.2.3.3   Summary of Comparison of
                  Perspectives of Mental Models 30
2.3   Application of Adaptable Framework:
      Charactering Mental Models by Perspective
      and Evaluating 'Risk of Bias' 31
2.4   Conclusions 32

Chapter 3   The *Quick Association Check* (QuACk): A Resource-Light,
                'Bias Robust' Method for Exploring the Relationship
                between Mental Models and Behaviour Patterns with
                Home-Heating Systems 37

3.1   Introduction 37
3.2   Methods Used for the Development and
      Evaluation of QuACk 39
      3.2.1   Literature Review 40
      3.2.2   Assess Methods for Home-Heating Context 41
              3.2.2.1   Content Analysis of Previous Research 47
      3.2.3   Existing Categories of Mental Models and
              Behaviour 47
      3.2.4   Consider Bias in Mental Models Research 49
      3.2.5   Developing Data Collection Method 49
              3.2.5.1   Paper-Based Activities 51
              3.2.5.2   Verification of Outputs 51
      3.2.6   Developing Analysis Method 52
3.3   Pilot Case Studies and Participant Observation for
      Data Collection 53
3.4   Participant Observation – Data Analysis 55
      3.4.1   Applying the Analysis Reference Table 55
              3.4.1.1   Behaviour Pattern 55
              3.4.1.2   Mental Model Description of
                        Home-Heating Function 61
      3.4.2   Benefits of Output Formats 64
              3.4.2.1   Self-Report Diagram 64
              3.4.2.2   Mental Model Description 64
              3.4.2.3   Association between Mental Model of
                        Device Function and Behaviour 64
      3.4.3   Evaluating the Utility of the Analysis
              Reference Table 65
      3.4.4   Improvements to the Analysis Reference Table 65

3.5   Validation...67
        3.5.1   Measurement Validity of Self-Report
                 Behaviour...67
        3.5.2   Reliability of Analysis Method.................................69
                 3.5.2.1   Dynamics of Exercise...............................69
                 3.5.2.2   Results of Inter-Analyst Reliability
                            Exercise..69
                 3.5.2.3   Improvements...71
3.6   Discussion...71
        3.6.1   Method Evaluation..73
3.7   Conclusions..75

Chapter 4   Case Studies of Mental Models in Home Heat Control:
                Searching for Feedback, Valve, Timer and Switch Theories.............77

4.1   Introduction..77
4.2   Method..80
        4.2.1   Participants and Setting...80
        4.2.2   Data Collection...81
        4.2.3   Dynamics of the Interview..82
        4.2.4   Analysis of Outputs..87
4.3   Case Studies..89
        4.3.1   Participant A: A Feedback Mental Model of
                 Thermostat with Elements of Valve
                 Behaviour...89
        4.3.2   Participant B: Feedback Behaviour without a
                 Feedback Mental Model...95
        4.3.3   Participant C: Timer Model for Alternate
                 Control Devices...99
4.4   Discussion...105
4.5   Conclusions...109

Chapter 5   When Energy-Saving Advice Leads to More, Rather
                Than Less, Consumption...111

5.1   Introduction...111
5.2   Method...115
        5.2.1   Participants...115
        5.2.2   Setting..116
        5.2.3   Data Collection...116
        5.2.4   From Central Heating System..................................116
        5.2.5   From the User..117
5.3   Results and Discussion...117
5.4   Summary and Conclusions..133

Chapter 6 Mind the Gap: A Case Study of the Gulf of Evaluation and
Execution of Home-Heating Systems 135

    6.1   Introduction 135
        6.1.1   Norman's (1986) Gulf of Evaluation and Execution..... 137
        6.1.2   Conceptual and Mental Models of Home-
                 Heating Systems 139
    6.2   The Design Model 140
        6.2.1   The Design Model Expressed as an Expert
                 'User Mental Model' 142
        6.2.2   What Does 'Appropriate' Home-Heating
                 Control Look Like? 145
        6.2.3   Stages of 'Appropriate' Activity with a
                 Home-Heating System 145
    6.3   System Image of Home Heating... 150
        6.3.1   Home Heating at the 'System' Level 150
        6.3.2   Home Heating at the Device Level........................... 152
    6.4   The User's Mental Model of Home Heating – Case Study
        Results and Discussion 154
        6.4.1   How Compatible Were the Case Study User
                 Mental Models of Home Heating? 155
        6.4.2   How Appropriate Were Case Study Self-Reported
                 Behaviour of Home-Heating Operation? 155
        6.4.3   A Discussion of the Seven Stages of Activity When
                 Users Operate Their Home-Heating System.............. 162
                 6.4.3.1   The Gulf of Execution 162
                 6.4.3.2   The Gulf of Evaluation 167
    6.5   Conclusions 169

Chapter 7 Using Interface Design to Promote a Compatible User Mental
Model of Home Heating and Pilot of Experiment to Test the
Resulting Design 183

    7.1   Introduction 183
    7.2   Concept Development 186
        7.2.1   Design of Key Devices 186
                 7.2.1.1   Thermostat 186
                 7.2.1.2   Programmer 188
                 7.2.1.3   Boost 190
                 7.2.1.4   TRV 191
        7.2.2   Design of System View 192
        7.2.3   Creating a Simulation... 195
        7.2.4   Pilot 197
    7.3   Discussion... 197
    7.4   Conclusion.. 200

Chapter 8 Mental Model Interface Design: Putting Users in Control
of Their Home-Heating Systems 201

8.1   Introduction 201
8.2   Method...204
       8.2.1   Experimental Design 205
       8.2.2   Participants 207
       8.2.3   Apparatus and Materials...207
       8.2.4   Procedure...208
8.3   Results 209
       8.3.1   User Mental Models of Home-Heating
                Simulation...209
       8.3.2   User Behaviour with Home-Heating
                Simulation 215
                8.3.2.1   Underlying Assumption for Study...........215
8.4   Discussion..221
       8.4.1   Improved Discoverability of Home-Heating
                Controls 221
       8.4.2   More Appropriate Mental Models............................222
       8.4.3   Increased Use of Frost Protection and
                Holiday Buttons...223
       8.4.4   More Appropriate Behaviour with
                TRV Controls 223
       8.4.5   Greater Control of Boiler Activation........................224
       8.4.6   Increased Goal Achievement....................................225
                8.4.6.1   Limitations of Study 225
8.5   Conclusions 226

Chapter 9 Conclusion..227

9.1   Introduction 227
9.2   Summary of Findings 227
       9.2.1   Bias Must Be Considered in Mental
                Models Research 228
       9.2.2   Outputs from QuACk Help Explain Energy-
                Consuming Behaviour..228
       9.2.3   We Need to Think beyond the Thermostat –
                Home-Heating Behaviour Should Be Understood
                at a System Level...229
       9.2.4   Broader System Variables Need to Be
                Understood for Optimal Consumption, but Are
                Not Promoted by Existing Technology 229
       9.2.5   Mental-Model-Driven Design Helps Users
                Achieve More Heating Goals 230

9.3   Core Issues Relating Particularly to the Home-Heating
       Case Study 230
       9.3.1   Optimal Home Heat Control Is a Complex Task 230
       9.3.2   Existing Technology Does Not Support
                a 'Systems UMM' of Home Heating 231
       9.3.3   We Cannot Control All the Variables That Affect
                Optimal Home Heating Control 232
9.4   Recommendations 232
       9.4.1   Recognize the Complexity of the Task for
                Householders, When Embarking on Strategies to
                Reduce Home-Heating Consumption 232
       9.4.2   Use System-Level Strategies for Encouraging
                Appropriate Home-Heating Consumption 233
       9.4.3   Use a Mental Models Approach When Seeking
                to Encourage Appropriate Behaviour in Complex
                Systems 233
       9.4.4   Design Future Heating Systems with Optimal
                Consumption as the Primary Goal 233
9.5   Areas of Future Research 234
       9.5.1   Extension of the 'Tree-Ring' Method for
                Considering Bias 234
       9.5.2   Extension of the QuACk Method for Exploring
                Association with Mental Models and Behaviour 234
       9.5.3   Tailored Guidance for Optimal Home-Heating
                Behaviour in Different Circumstances 234
       9.5.4   Enhancement to Home-Heating Control Panel
                and Testing in Domestic Setting 234
9.6   Concluding Remarks 235

Appendix A: The Quick Association Check 237
A.1   QuACk Instructions for Interviewer 237
       A.1.1   Background 237
       A.1.2   Preparation 238
                 A.1.2.1   Provide Participant Information Sheet 238
                 A.1.2.2   Verbal Positioning 238
       A.1.3   What to Expect and How to Deal with It 238
       A.1.4   Interview Outputs 239
A.2   QuACk Participant Information Sheet 239
A.3   QuACk Interview Template for Home-Heating Domain 240
       A.3.1   Background Experience in Home Heating 240

A.3.2 Behaviour 241
        A.3.2.1   Self-Report on Usage 242
        A.3.2.2   Response to Scenarios 243
A.3.3 Mental Model of Device Function 244

Appendix B 247

References 267

Bibliography 277

Index 279

List of Figures

Figure 1.1   Relationship between hypotheses. 3

Figure 2.1   Illustrating the way different theorists place mental models in
relation to other knowledge structures. 14

Figure 2.2   Depicting how information in the world can be filtered in
different ways. 16

Figure 2.3   Providing an analogy for 'Bias rings' 17

Figure 2.4   Proposing the interaction between bias rings and filters 18

Figure 2.5   Depiction of the relationship between agents and models
when identifying a shared theory of the operation of a home
thermostat (based on Kempton's [1986] study). 20

Figure 2.6   The influence of 'background bias' in forming an agent's
mental model is shown as a patterned ring around each agent. 21

Figure 2.7   The influence of social bias on forming an agent's
mental model. 22

Figure 2.8   Tree-ring profiles showing the layers of bias that alter the
construction and access of an agent's mental model. 23

Figure 2.9   Norman's (1983) definitions of mental models represented
schematically. 24

Figure 2.10 Tree-ring models representing the bias explicit in Norman's
(1983) definitions. 25

Figure 2.11 Re-representation of Kempton's (1986) subject's individual
mental model, in light of Norman's (1983) definition. 26

Figure 2.12 Wilson and Rutherford's (1989) definitions of mental model
concepts, depicted as a schematic and tree-ring profile. 27

Figure 2.13 Four different schematics for Wilson and Rutherford's (1989)
definition of a 'user's mental model'. 28

Figure 2.14 Tree-ring models representing the four different schematics
for Wilson and Rutherford's (1989) definition of a 'user's
mental model', highlighting the variation in biases based on
interpretation. 29

Figure 2.15 Re-representation of Kempton's (1986) analyst's shared theory,
in light of Wilson and Rutherford's (1989) User Mental Model
definition (assuming contact with the user). 30

Figure 2.16 Comparing Kempton's (1986) 'shared theory' of thermostat function with Payne's (1991) 'mental model' of bank machine function, showing model source (subjects), intermediary (analyst) and recipient (academic community)..............................32

Figure 3.1 Process for method development of the quick association check for home heating (QuACk)..40

Figure 3.2 Components of QuACk prototype including elements informed by Payne (1991), Oppenheim (1992), Kempton (1986) and Revell and Stanton (2012). 50

Figure 3.3 Example of QuACk outputs: (a) template with annotated self-report of home-heating use; (b) user mental model description of home-heating function..51

Figure 3.4 The key elements of QuACk following iterative development.........57

Figure 3.5 Output 1 from QuACk, redrawn for clarity, showing 'behaviour when using home heating over a typical week'...............................60

Figure 3.6 Mental model description of device function for participant 3, redrawn from output 2 for clarity...62

Figure 4.1 The layout of the home-heating devices and specific models used during the inter-group case study. 81

Figure 4.2 Participant A's output from the paper-based activity, which formed the basis for analysis. 86

Figure 4.3 User-verified mental model description of home heating for participant A..90

Figure 4.4 Isolated causal relationship for thermostat knob taken from participant A's user-verified mental model description of home heating...92

Figure 4.5 User-verified mental model of home heating for participant B. 96

Figure 4.6 Isolated causal relationship for thermostat set point taken from participant B's user-verified mental model description of home heating...97

Figure 4.7 User-verified mental model description of home-heating function, from participant C. 100

Figure 4.8 Cause and effect route of schedule function for participant C. 102

Figure 4.9 Cause and effect depiction of override button for participant C.....103

Figure 5.1 Graph to compare boiler 'on' periods for three matched households over a single week during winter in the United Kingdom. 112

Figure 5.2   Remotely recorded thermostat set points, internal temperatures
and boiler 'on' periods during a single week for house X............. 119

Figure 5.3   Remotely recorded thermostat set points, internal temperatures
and boiler 'on' periods during a single week for house Y............. 120

Figure 5.4   Remotely recorded thermostat set points, internal temperatures
and boiler 'on' periods during a single week for house Z. 121

Figure 5.5   Devices used and typical adjustments made over a typical
week reported by participant X.. 122

Figure 5.6   Devices used and typical adjustments made over a typical
week reported by participant Y. 123

Figure 5.7   Devices used and typical adjustments made over a typical
week reported by participant Z... 124

Figure 5.8   User mental model description of the home-heating system
for participant X... 127

Figure 5.9   User mental model description of the home-heating system
for participant Y. 128

Figure 5.10   User mental model description of the home-heating system
for participant Z. 129

Figure 6.1   Norman's (1986) seven stages of user activity applied to
home-heating context.. 138

Figure 6.2   According to Norman (1986), the system image contributes to
the user's mental model, influencing their interaction with the
heating system... 140

Figure 6.3   The compatibility of the user's mental model to the design
model at each of the seven stages of activity characterize
the 'structural integrity' of the bridges that span the gulf of
evaluation and execution... 141

Figure 6.4   The home-heating system 'design model' represented as an
expert user mental model description. 143

Figure 6.5   The elements of the design model that should be evident in
compatible user mental models of a home-heating system. 144

Figure 6.6   Recommended state of the home-heating system – stage 3 of
Norman's seven stages of activity... 146

Figure 6.7   Norman's (1986) seven stages of activity broken down by
typical home-heating goals... 147

Figure 6.8   The 'system image' of the home-heating system, showing the
layout and device interface of the home-heating elements from
the 'compatible mental model'. 150

Figure 6.9   Key elements compatible to design model, for each
                  participant – greyed-out areas are missing elements. 156

Figure 6.10   Figure to compare the elements of householders' user mental
                  models with those in the proposed 'compatible' model. 159

Figure 6.11   Householders' self-report of typical use of home-heating
                  controls over a week period. 160

Figure 7.1   'Realistic' style thermostat interface with a 'flame' icon
                  indicating boiler operation and ambiguous label 'room' to
                  identify where temperature samples are fed back to the device. 187

Figure 7.2   Redesign of thermostat interface to promote appropriate device
                  model to users. 187

Figure 7.3   'Realistic' home-heating programmer interface. 189

Figure 7.4   Redesign of programmer to simplify schedule input, to
                  encourage inclusion in behaviour strategies. 189

Figure 7.5   'Realistic' boost button, as a feature on a programmer device. 190

Figure 7.6   Redesign of boost button to promote 1-hour operation. 190

Figure 7.7   'Realistic' style of TRV control, with ambiguous 5-point scale 191

Figure 7.8   Redesign of TRV controls to promote heat-limiting feedback
                  device model. 192

Figure 7.9   Distribution of typical home-heating controls across the home
                  (but users can typically only see one room at a time) 193

Figure 7.10   Redesign of interface to display distributed control devices
                  with cause-and-effect links. 194

Figure 7.11   Control panel for key controls emphasizing link between set
                  point choice with key controls and boiler status. 194

Figure 7.12   Interface created to represent a typical home-heating
                  interaction. 196

Figure 7.13   Interface designed to promote compatible user mental model of
                  home-heating system. 196

Figure 8.1   Different variables that effect home-heating behaviour and its
                  consequences. 203

Figure 8.2   Diagram to show relationship between theses hypotheses and
                  specific hypotheses tested in the thesis. 205

Figure 8.3   Screen shots of realistic interface (a) and design interface (b). 206

Figure 8.4   Range of heating controls found in user mental models 210

Figure 8.5   Frequency of participants who described key heating controls
in user mental model descriptions. 211

Figure 8.6   Frequency of participants who described advanced heating
controls in user mental model description. 211

Figure 8.7   Frequency of appropriate functional models for key controls. 212

Figure 8.8   Graph to compare the frequency of appropriate and
inappropriate functions assigned to key controls. 212

Figure 8.9   Number of key system elements present in UMM descriptions. 213

Figure 8.10   Frequency of key system elements present in UMM
descriptions. 214

Figure 8.11   Proportion of controls used in simulation, depending on
presence in UMM. 216

Figure 8.12   The frequency of use for controls. 216

Figure 8.13   Frequency of TRV set point adjustments. 217

Figure 8.14   Mean range of TRV set point values. 218

Figure 8.15   Frequency of use and set point choice over time of TRVs. 219

Figure 8.16   Control of boiler activation by thermostat adjustments. 220

Figure 8.17   Percentage of thermostat set point choices leading to boiler
state change. 220

Figure 8.18   Total proportion of time within goal temperature range. 221

List of Tables

Table 2.1 Categorise and Compare the Types of Knowledge Structures Proposed by Johnson-Laird (1983), Moray (1990) and Bainbridge (1992).. 11

Table 2.2 Application of Proposed Framework to Kempton (1986) and Payne (1991) to Better Specify Mental Models for Commensurability, Considering the Risk of Bias in Interpretation and the Perspective from Which Data Are Gathered............................ 33

Table 3.1 Methods for Identifying Mental Models Associated with Behaviour by Rouse and Morris (1986), Evaluated for Domestic Home-Heating Context, Based on Speed, Ease and Cost of Data Collection and Analysis.. 42

Table 3.2 Subject and Purpose of Probes Used by Kempton (1986) and Payne (1991).. 48

Table 3.3 Description of Mental Model Categories of Home Heating from the Literature 52

Table 3.4 Bias Rings Identified in the Collection and Analysis of Data Derived from Interviews, Their Cause and the Mitigation Strategy Employed in the Development of QuACk Prototype 54

Table 3.5 Analysis Reference Table for Quick and Systematic Analysis of Outputs from QuACk 56

Table 3.6 Iterations to QuACk Resulting from Case Study and Participant Observations (Round Bullets Reflect Amendments to Method, Dash Bullets Identify Aspects that Worked Well) 58

Table 3.7 Summary of Evaluation of Analysis Reference Table........................ 66

Table 3.8 Summary of Spouses' Agreement with Behaviour Shown in Output 1 (Agreement = ✓, Disagreement = ✗).. 68

Table 3.9 Results of Inter-Analyst Reliability Exercise 70

Table 4.1 Risk of Bias and Mitigation Strategy for Method Adopted, Derived Using Tree-Ring Method from Revell and Stanton (2012) as Described in Chapter 2 83

Table 4.2 Analysis Table for Categorizing Responses from Interview Transcripts 88

Table 6.1 Expert Considered 'Essential' Components of Compatible User Mental Model for Appropriate Operation of Heating System............ 144

Table 6.2 Summary of Analysis of the System Image of the Heating System, with Possible Misunderstanding by the User 151

Table 6.3 Comparison of Householders' Actions to Achieve Goals, with Expert Recommendations 163

Table 6.4 Summary of Perceptual Cues Used by Participants to Evaluate the State of the System 168

Table 6.5 How the System Image of the Home-Heating System Can Effect User Mental Models That Underpin Norman's (1986) Seven Stages of Action 170

Table 7.1 Summary of Issues and Changes to Experimental Procedure, Interfaces and Setting, Following Pilot Run 198

Table B.1 Part 1 – Output 1 Analysis Table for Categorizing Behaviour Patterns of Home Heating 248

Table B.2 Part 2 – Output 2 Analysis Table for Categorizing Mental Model Descriptions of Home Heating 256

Table B.3 Part 3 – Walk-Through Questions to Guide Analysts When Categorizing Output 1 from QuACk 262

Table B.4 Part 4 – Walk-Through Questions to Guide Analysts When Categorizing Output 2 from QuACk 264

Preface

This book has arisen from the desire to demonstrate how the somewhat intangible notion of 'mental models' can be used as a design tool to improve user interaction. Interest was ignited in this area more than 30 years ago, when two very different seminal books titled *Mental Models* appeared in 1983. One, written by Philip Johnson-Laird, considered this knowledge construct from a psychological and linguistic point of view. The other, a collection of papers edited by Dedre Gentner and Albert L. Stevens, provided a more experimental and practical approach to understanding the notion. Both, however, looked to understanding mental models as a mechanism to predict and explain human comprehension and behaviour. It was Willet Kempton who, just a few years later in 1986, expanded this notion to groups of people, rather than individuals, in the guise of a 'folk theory'. He applied mental models to a problem that 30 years later still remains of global concern, that of energy conservation. Kempton pointed to the role of design in developing appropriate users' mental models, complementing factors such as training and experience (that had been the focus of Johnson-Laird and Gentner and Stevens).

Mental models have face validity that is attractive to designers and interface developers, and frequently feature as a strategy for good interface design (e.g. in the Mac operating system human interface guidelines). Producing a mental-model-inspired design can be easier said than done, however, and this book aims to demonstrate the challenges and considerations faced by a designer or practitioner looking to guide human behaviour. In particular, we hope that this book makes a clear case for a broader view than the interaction between user and control, when designing interactions with domestic energy systems.

The research presented in this book was undertaken as part of a broader project exploring the utility of intelligent agents for home energy management sponsored by the Engineering and Physical Sciences Research Council in the United Kingdom, resulting from collaboration between the Sustainable Energy Research Group, the School of Electronics and Computer Science, and the Transportation Research Group, within the University of Southampton. Neville Stanton has a huge body of work relating to knowledge constructs in multiple domains, covering not only mental models but also schema theory. Kirsten Revell draws together her experience in psychology and industrial design to offer a pragmatic approach to interaction design.

This book will bring the reader on the journey the authors have travelled, inspired by Kempton's original insights that influencing householders' mental models could result in large-scale reductions in energy consumption. The focus of the book is on making explicit the implicit, not only in terms of representing user mental models, but also the methodology and approach taken to produce the designs.

Acknowledgements

Financial support for the research presented in this book was provided by the Engineering and Physical Sciences Research Council (EPSRC) as part of a broader project 'Intelligent Agents for Home Energy Management' (IAHEM), a collaboration between the School of Electronics and Computer Science, Sustainable Energy Research Group, and Transportation Research Group at the University of Southampton. We extend thanks to all those who contributed to the research opportunities that made this book possible, in particular Professor Alex Rogers, Professor AbuBakr Bahaj, Dr. Patrick James, Sid Ghost, Rama Kota, Andrei Petre, David Podesta and expert help from Horstmann Ltd.

Authors

Kirsten Revell, PhD, is a human factors engineering research fellow in the Transportation Research Group in the Faculty of Engineering and the Environment at the University of Southampton. Dr Revell graduated from Exeter University in 1995 with a BSc (Hons) in psychology. After graduating, Kirsten spent six years working for Microsoft Ltd., managing the EMEA Services Academy, before undertaking a second degree in industrial design, at Brunel University. During an internship with the Ergonomics Research Group at Brunel, she joined a major field trial for the Human Factors Integration Defence Technology Centre (HFI DTC) focused on the usability of digital mission planning and battle-space management systems. She brought together both psychology and design disciplines during an EPSRC-sponsored PhD in human factors at the University of Southampton, investigating a mental models approach to behaviour change with domestic energy systems. More recently, Dr Revell worked on the Innovate UK Future Flight Deck Technologies project, collaborating with BAE systems, General Electric and Coventry University, to evaluate how technology could enable changes to the design of the flight deck and crewing configurations. She is now investigating human interaction in the design of autonomous vehicles (Hi:Dav) in a JLR/EPSRC-funded project collaborating with the Cambridge University. Dr Revell's research passion is to understand how the design of tools, interfaces and systems impacts user capabilities and behaviour.

Professor Neville Stanton, PhD, DSc, is a chartered psychologist, chartered ergonomist and chartered engineer. He holds the chair in Human Factors Engineering in the Faculty of Engineering and the Environment at the University of Southampton in the United Kingdom. He has degrees in occupational psychology, applied psychology and human factors engineering, and has worked at the Universities of Aston, Brunel, Cornell and MIT. His research interests include modelling, predicting, analysing and evaluating human performance in systems, as well as designing the interfaces and interaction between humans and technology. Professor Stanton has worked on the design of automobiles, aircraft, ships and control rooms over the past 30 years, on a variety of automation projects. He has published 35 books and more than 270 journal papers on ergonomics and human factors. In 1998, he was presented with the Institution of Electrical Engineers Divisional Premium Award for research in system safety. The Institute of Ergonomics and Human Factors in the United Kingdom awarded him The Otto Edholm Medal in 2001, The President's Medal in 2008 and The Sir Frederic Bartlett Medal in 2012 for his contributions to basic and applied ergonomics research. The Royal Aeronautical Society awarded him and his colleagues the Hodgson Prize in 2006 for research on design-induced, flight-deck error published in *The Aeronautical Journal*. The University of Southampton has awarded him a Doctor of Science in 2014 for his sustained contribution to the development and validation of human factors methods.

List of Abbreviations

C(M(t))	Scientists' conceptualisation of users mental model of target system
C(t)	Conceptual model of target system
CUMM	Compatible user mental model
DCM	Designer's conceptual model
ECL	Efficient cause lattice
EPSRC	Engineering & physical sciences research council
FCL	Formal cause lattice
FiCL	Final cause lattice
HCI	Human–computer interaction
MaCL	Material cause lattice
MCL	Mental cause lattice
MM	Mental model
M(t)	Users mental model of target system
PSL	Physical system lattice
QuACk	Quick association check
t	Target system
TRV	Thermostatic radiator valve
UCM	User's conceptual model
UMM	User mental model

1 Introduction

1.1 WHO IS THIS BOOK FOR?

This book is for anyone interested in understanding how and why people interact with domestic energy systems, or who want to design controls or systems that encourage sustainable behaviour. It demonstrates why the answer to reducing domestic energy consumption isn't just to 'turn down the thermostat' and offer insights and practical guidance for a wide range of readers. For those from the human factors engineering field, this work is heavily grounded in the literature in the field, extending theory, methodology and application with in-depth worked case studies. Designers and electrical engineers involved in the technical and functional design of heating controls would also gain value in understanding why users behave unexpectedly when interacting with technology and how this knowledge can drive better interface design (Chapters 4, 6 and 7). Professionals in the built environment with an interest in sustainability or end-user energy consumption would also benefit from reading this book to gain a broader appreciation of the cognitive and behavioural variables that interact with structural considerations to determine consumption.

1.2 HOW DIFFERENT READERS SHOULD APPROACH READING THE BOOK?

Readers with an academic bias will appreciate that each chapter references a large body of literature, uses an evidence based approach to forming conclusions, and provides extensive references at the end of the book for further reading. Those with a more pragmatic focus can skim through the theoretical introductions to each chapter and focus on worked examples in the main body. All readers will appreciate the extensive use of diagrams, to make abstract concepts more 'concrete', and the examples provided of data, outputs, processes and designs. Readers interested in full-colour versions of design examples will be able to access these online through the publisher website.

1.3 BACKGROUND

The United Kingdom has legislated to cut greenhouse gas emissions by 80% by 2050 (Climate Change Act 2008). A key element in achieving the proposed reduction in CO_2 emissions is the need to support domestic consumers in both reducing their demand for energy and improving the efficiency with which they use it. These consumers currently have the least visibility regarding their energy use, but they collectively contribute over 25% of total UK carbon emissions (The UK Low Carbon Transition Plan 2009). A significant contribution to the variability of domestic energy use across buildings is due to behavioural differences of the householders

(Lutzenhiser and Bender 2008). The work in this book starts from the premise it is people, not houses, that use energy. How and why people use energy, whether they are using it effectively, and how interface design can form a key part of strategies to reduce consumption is the impetus of this work.

Since home heating is responsible for 58% of domestic energy use, the key focus was on users' mental models (UMMs) of a domestic heating system. The heating system is used as a case study throughout the book to demonstrate in depth how application of a mental model approach to design can, and should, be adopted in complex systems such as the domestic setting. The approach used can be adopted for other domestic energy systems where the householder is responsible for controlling operation, particularly if the function of the system is not explicit. In particular, the insights and findings in this book are equally relevant for cooling systems (e.g. air conditioning) whose controls have a similar function to those operating heating systems.

Mental models are thought to be representations of the physical world (Veldhuyzen and Stassen 1976, Johnson-Laird 1983, Rasmussen 1983), constructs that can explain human behaviour (Wickens 1984, Kempton 1986) and internal mechanisms allowing users to understand, explain, operate and predict the states of systems (Craik 1943, Kieras and Bovair 1984, Rouse and Morris 1986, Hanisch et al. 1991). The notion of mental models has proven attractive for many domains: for psychology, when considering cognitive processing (Johnson-Laird 1983, Bainbridge 1992); in interface design (Williges 1987, Norman 2002, Jenkins et al. 2010); to promote usability (Norman 2002, Mack and Sharples 2009, Jenkins et al. 2011); and for the human factors domain, to enhance performance (Stanton and Young 2005, Stanton and Baber 2008, Grote et al. 2010, Bourbousson et al. 2011) and reduce error (Moray 1990, Stanton and Baber 2008, Rafferty et al. 2010). A mental models approach can therefore tackle behaviour change from a variety of perspectives, extending the reach of the findings.

1.4   AIMS AND OBJECTIVES/PURPOSE

This research undertaken focused on how the concept of mental models can be applied in design to elicit behaviour change that results in increased achievement of home-heating goals (such as reduced waste and improved comfort). This research also contributes to methodology regarding the extraction and application of mental models, allowing the notion of mental models to be applied across domains when behaviour change was sought. The research presented in this book can be broken down into three sub-hypotheses on which the overall hypothesis depends.

1.4.1   OVERALL HYPOTHESES

The following hypothesis encompasses the focus for this book:

> The knowledge of existing mental models of devices can be used in device design to encourage patterns of device use that increase goal achievement (e.g. balance between comfort and consumption).

1.4.2 Sub-Hypotheses

The overall hypothesis can be broken down into the following sub-hypotheses, which will be explored in different chapters in this book:

Hypothesis 1: 'Users' mental models of devices influence their pattern of device use'.

Hypothesis 2: 'Patterns of device use influence the amount of energy consumed over time'.

Hypothesis 3: 'The design of device influences the mental models of those devices'.

The relationship between these hypotheses is represented in Figure 1.1, which is informed by the work of Kempton (1986). The research assumes that a causal relationship can be found between the different components (energy-consuming

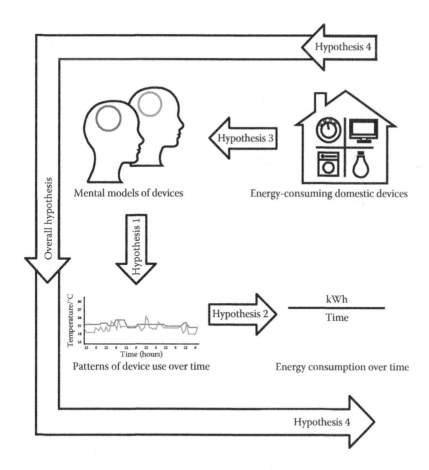

FIGURE 1.1 Relationship between hypotheses.

domestic devices, mental models of devices, patterns of device use and energy consumption over time) and that changes in the design of devices can instigate changes in behaviour.

1.5   CONTRIBUTION TO RESEARCH

This book makes a contribution to the existing body of research in academic literature in three key ways: First, it provides generic methodologies relating to the investigation and application of mental models in the form of the tree-rings framework, to consider layers of bias to encourage commensurability of findings (Chapter 2), and the application of Norman's Gulf of Evaluation and Execution (Chapter 6) at the 'mental model' level to aid design strategies. Second, it has developed (Chapter 3) and applied (Chapters 4 through 6) a home-heating-specific methodology for capturing and analysing mental model descriptions and associated behaviour, at device and system levels. Finally, home-heating-specific findings that further the body of knowledge have been identified: specifically that system-level analysis is necessary for behaviour change strategies (Chapters 3 through 5 and 6); that gaps in user mental models go a considerable way in explaining energy-consuming behaviour (Chapters 4 and 5); and importantly, that interface design can alter the mental model held, *without prior training*, to affect users' behaviour with controls (Chapter 8). The implications of these findings are far-reaching, not only when considering potential domestic energy savings to mitigate climate change, but for any domain where technology has been designed for performance goals, yet requires active control by users, as well as understanding of broader variables, to fulfil conservation and efficiency goals.

1.6   STRUCTURE OF THE BOOK

This book has been written to provide flexibility to the reader, based on their interest and needs. It is organized in nine chapters, each building on information and insights gained from previous chapters. As such it can be read from start to finish, whereby the reader can see the journey the authors have made; firstly, in terms of their understanding of their applicability of mental model theory to design, and, secondly (and more significantly), the move from considering interaction and design at the device level, focused purely on the thermostat control, to the system level. However, for readers who wish to dip into specific areas, each chapter can also be read in isolation. For this flexibility to be possible, it necessitated key concepts relating to mental models theory to be repeated in each chapter so the positioning and implications of the insights were clear.

An overview of the chapters follows.

Chapter 1 – Introduction
    This chapter introduces the background and impetus to this work. It outlines
        the main research objectives, provides guidance to the reader as to what to
        expect, including a summary of each chapter.

Chapter 2 – Models of Models: Filtering and Bias Rings in Depiction of Knowledge Structures and Their Implications for Design

This chapter investigates the barriers to be overcome in order to apply the notion of mental models pragmatically. Literature from psychology, human–computer interaction (HCI) and human factors sources were reviewed to determine the utility of 'mental models' as a design tool. The chapter identifies bias as a major impediment to pragmatic application and concludes that definition and methods of construction and access need to be sufficiently specified. The chapter develops a graphical method to compare existing research in mental models, highlighting similarities, differences and ambiguities. This 'tree-rings' method is applied to the types of mental models described in the work of Kempton (1986) and Payne (1991) to illustrate fundamental differences in the notion.

Chapter 3 – The Quick Association Check *(QuACk): A Resource-Light, 'Bias Robust' Method for Exploring the Relationship between Mental Models and Behaviour Patterns with Home Heating Systems*

Simple application methods, which allow exploration of a link between users' mental models of a device and their behaviour with that device, are scarce in the literature. This chapter describes the development of the Quick Association Check (QuACk) – a semi-structured interview with paper-based activities and templates. QuACk collects data, verified by the user, relating to: (1) typical behaviour patterns when operating home heating, and (2) mental model descriptions of home-heating function. The aim of QuACk was to produce a quick, resource-light method to explore the association between mental models and behaviour patterns with home heating to support studies targeted at hypothesis 1 (Figure 1.1). QuACk was developed with consideration of bias from the outset, using the tree-ring method described in Chapter 2. The outputs from a single case study are used to illustrate the method and the process of analysis and the potential ways that QuACk could inform energy-reducing strategies are discussed.

Chapter 4 – Case Studies of Mental Models in Home Heat Control: Searching for Feedback, Valve, Timer and Switch Theories

This chapter was inspired by the work of Kempton (1986), who identified two common mental models held by householders of the home-heating thermostat: (1) a 'feedback' model that was a simplified but correct version of the way the device worked that could lead to energy systematically wasted, and (2) a 'valve' model that misunderstood the way the device functioned, but could result in energy saving. An intergroup case study is presented that investigated present-day mental models of thermostat function that differ significantly from actual functioning. These models were categorized according to Kempton's (1986) valve and feedback shared theories, and others from the literature. Distinct, inaccurate mental models of the heating system, as well as thermostat devices in isolation, are described, and their relationship to self-reported behaviour is reported in support of hypothesis 1 (Figure 1.1). The chapter highlights the need to consider the mental models of the heating system in terms of an integrated set of control devices, and to consider users' goals and expectations of the system benefit.

Chapter 5 – When Energy Saving Advice Leads to More, Rather Than Less, Consumption

Where Chapter 4 explored differences in mental models of the thermostat, Chapter 5 considers householders that share the same model. A case study of three households that held a 'feedback' mental model of the home-heating thermostat, as defined by Kempton (1986), was undertaken to understand the driver behind differences in their home-heating strategies, and the effect on energy consumption. The chapter provides evidence in support of hypotheses 1 and 2 (Figure 1.1). Five different data sources were used for analysis, comprising: (1) 'boiler on' durations, (2) thermostat set point adjustments, (3) self-reported strategies with home-heating controls, (4) user mental model descriptions of the home-heating system and (5) interview transcripts. The chapter found that differences in user mental models of home heating at the system level explained the differences in the strategies chosen at the control device level, building on the conclusions from Chapter 3. Differences in 'boiler on' periods were found to relate to limitations of Kempton's (1986) 'feedback' mental model. The implications for energy-consuming strategies are discussed.

Chapter 6 – Mind the Gap: A Case Study of the Gulf of Evaluation and Execution of Home-Heating Systems

This chapter applies Norman's (1983) idea of the 'gulf of evaluation and execution' to the home-heating domain. The mental model from a home-heating expert is used as a representation of the 'design model' of the system. How the mental models of novice home-heating users differ from this design model represents Norman's (1986) gulf. A design specification is developed based on common omissions and misunderstandings found in novice users. The chapter highlights how broader variables (missing from a feedback mental model), such as household thermodynamics, would facilitate a user mental model of home heating that enables appropriate behaviour with controls. How a typical home-heating interface at a system and device level impedes a compatible user mental model to the design model is explored to identify where design strategies can help bridge this gulf. The chapter provides a methodology to facilitate investigation of hypotheses 3 and 4 (Figure 1.1).

Chapter 7 – Using Interface Design to Promote a Compatible User Mental Model of Home Heating and Pilot of Experiment to Test the Resulting Design

This chapter builds on Chapter 6 by taking the design specification produced to bridge the gulf of evaluation and execution experienced by domestic users of heating systems and applying it to the design of a home-heating control panel. The chapter uses design principles recommended by Norman (2002) and Manktelow and Jones (1987) to evoke mental models in the user. The aim was to create a 'mental model promoting interface' that could form part of an experiment that compared performance, behaviour and models evoked, with a more traditional interface, to test hypotheses 3 and 4 (Figure 1.1). The focus was on key devices (Thermostat, Programmer, Thermostatic Radiator Valves [TRVs] and Boost Button) and their relationship with boiler activation and radiator output. This chapter shows concept developments for the redesign of key devices to promote appropriate functional models. At the system level, a redesign of the layout of key devices is shown to promote a mental model with appropriate integration between devices. The results of changes to the design of the control panel interface are shown following a pilot of the experiment.

Chapter 8 – Mental Model Interface Design: Putting Users in Control of Their Home-Heating Systems

This chapter reports the design and results of an experiment using a home-heating simulation to test hypothesis 4, that interface design can influence the achievement of home-heating goals by encouraging appropriate behaviour through the evocation of appropriate user mental models. Using the design concepts developed in Chapter 7, two interfaces were developed to compare a mental model promoting 'control panel' with a more traditional home-heating setup. The chapter reveals the benefits of a mental-model-driven design in terms of home-heating goal achievement, behaviour with heating controls and evocation of appropriate mental models at the device and system levels. The chapter also discusses limitations of the study.

Chapter 9 – Conclusions

This chapter summarizes the key findings from this research, followed by a discussion of the core issues underlying the research. Key recommendations are made and areas for future work are presented.

2 Models of Models

Filtering and Bias Rings in Depiction of Knowledge Structures and Their Implications for Design

2.1 INTRODUCTION

This chapter focuses on the nature of the key concept of this chapter, the notion of the 'mental model' itself. It looks at the characteristics of mental models as described in the literature, and considers how the choice of methodology and perspective of definition affect confidence in the validity of the captured knowledge structure. The insights gained in this chapter inform the approach taken to test hypothesis 1 and 3 described in Chapter 1. To determine if users' mental models of devices influence their pattern of device use (hypothesis 1), it is necessary to understand what constitutes a 'mental model' associated with behaviour. To determine if device design influences mental models of devices (hypothesis 3), it is necessary to understand the mechanisms by which mental models are formed, and the barriers to access that enable changes in mental models to be observed.

Mental models are thought to be representations of the physical world (Veldhuyzen and Stassen 1976, Johnson-Laird 1983, Rasmussen 1983), constructs that can explain human behaviour (Wickens 1984, Kempton 1986) and internal mechanisms allowing users to understand, explain, operate and predict the states of systems (Craik 1943, Gentner and Stevens 1983, Kieras and Bovair 1984, Rouse and Morris 1986, Hanisch et al. 1991).

The notion of mental models has proved attractive for many domains: for psychology, when considering cognitive processing (Johnson-Laird 1983, Bainbridge 1992); in interface design (Carroll and Olson 1987, Williges 1987, Norman 2002, Jenkins et al. 2010); to promote usability (Norman 2002, Mack and Sharples 2009, Jenkins et al. 2011); and for the human factors domain, to enhance performance (Stanton and Young 2005, Stanton and Baber 2008, Grote et al. 2010, Bourbousson et al. 2011) and reduce error (Moray 1990, Stanton and Baber 2008, Rafferty et al. 2010). For the successful application of this notion, clarity in its meaning and guidance in its use are vital. However, the lack of consensus in this field, brought to light by Wilson and Rutherford (1989) over 20 years ago, remains unresolved.

It is proposed that for successful application of the notion of mental models (e.g. in device design), the accurate capture of the model held by the agent of interest (e.g. device user) is key. In the literature, a pragmatic means of considering the deviation in accuracy from the model source to the recipient is absent. When comparing different notions of the term 'mental model', existing reviews of the literature have provided useful textual categorizations and high-level frameworks, which go some way towards the goal of commensurability (see Richardson and Ball [2009] for a comprehensive review). The literature does not sufficiently emphasize the relationship between the type of notion, the methodology of capture and the risk of bias.

This review, therefore, has two aims: first, to compare and contrast the work of major research figures in the field and, second, through the development of an adaptable framework, emphasize the importance of characterizing mental model content accurately based on (1) the risk of bias resulting from the methodology undertaken and (2) the perspective (in terms of model source and type) from which mental models are considered.

To achieve these aims, this chapter will first contrast the theories of Johnson-Laird (1983), Moray (1990) and Bainbridge (1992), who offer distinctly different concepts of mental models as inferred knowledge. The critical elements found will form the basis for an adaptable framework to demonstrate the second aim. Next, the role of cognitive bias in mental models research will be considered and the understanding, in this chapter, of 'filtering information', will be described with a view to further developing the framework using the case study of Kempton (1986). Following this, the different perspectives by which major researchers in the field define mental models, and the methodology with which they have undertaken research will be compared schematically, using the framework, to illustrate similarities and differences. Finally, the framework will be used to compare two notions of mental models offered by Kempton (1986) and Payne (1991) to illustrate how the schematics can be deconstructed to help specify both the perspective from which data are gathered and the risk of bias in interpretation of the source model.

This chapter considers mental models formed by an individual and accessed by another individual. As such, it does not address related notions such as 'shared mental models' or 'team mental models'. Brewer's (1987 c.f. Stanton [2000]) distinction that 'schemas are generic mental structures underlying knowledge and skill, whilst mental models are inferred representations of a specific state of affairs' may be helpful when reading this chapter.

2.1.1   The Concept of Mental Models as Inferred Knowledge in Cognitive Processing

This section will compare three different approaches to the role of mental models in cognitive processing, proposed by Johnson-Laird (1983), Bainbridge (1992) and Moray (1990). These theorists are chosen to show that despite considerable differences in approach, focus and context, the fundamental ideas are common. The aim of this section is to emphasize mental models as one of a range of mental constructs and highlight the importance of the role played by background knowledge.

The term 'knowledge structure' is taken from Wilson and Rutherford (1989) to describe the descriptions that analysts make of a user's understanding. A concise definition of a mental model is absent in the work under appraisal, with the notion

TABLE 2.1

Categorise and Compare the Types of Knowledge Structures Proposed by Johnson-Laird (1983), Moray (1990) and Bainbridge (1992)

| Theorist | Knowledge Structure | | |
	Inferred	Background	Other
Johnson-Laird (1983)	Mental model	World model	Propositional representations
Bainbridge (1992)	Working storage	Knowledge base	Meta-knowledge
Moray (1990)	Mental causal models	Physical systems lattice	n/a

conveyed by the theorists partly by comparison to alternate constructs. To aid the readers understanding, Table 2.1 summarizes the different categories of knowledge structures common to these theorists. The role and interaction of these structures will be expanded upon in the following section, considering each theorist in turn.

A summary comparing these ideas will then be undertaken, culminating in a graphical representation comparing these knowledge structures and related processes, which will form the basis of an adaptable framework that is built upon in part two of this chapter.

2.1.1.1 Johnson-Laird (1983)

Johnson-Laird (1983) took a linguistic approach to the study of mental models, to understand the role they played in inference and reasoning. Johnson-Laird (1983, 1989, 2005) rejected formal logic as the driver for reasoning, showing instead that the manipulation of mental models makes it possible to reason without logic.

Johnson-Laird (1983) presented a building block approach to knowledge structures. Information is initially encoded as 'propositional representations', which themselves do not allow the user to go beyond the data in the proposition (Manktelow and Jones 1987). Johnson-Laird (1983) introduced the concept of 'procedural semantics' as the mechanism which determined if a propositional representation would remain in that form, or be combined into the 'higher' structure of a 'mental model'. These, according to Johnson-Laird (1983, 1989, 2005), do allow users to go beyond the data, experiencing events by proxy to make inferences, predictions and, ultimately, decide what action to take as a result (Manktelow and Jones 1987).

Johnson-Laird (1983) explicitly described seven procedures which are required to create and test a mental model. All procedures state reference to a type of knowledge structure termed a 'model of the world' (Johnson-Laird 1983, 1989, 2005). Manktelow and Jones (1987) interpret the first five procedures as 'general'. These provide the function of determining prior experience of the representation, amending the world model with new information, combining previous separate models based on new interrelated information, and establishing the truth of a proposition. Manktelow and Jones (1987) interpret the final two procedures as recursive, with the function of checking back through the mental model with reference to the world model to see if it is flawed, amending or changing to a better model if necessary.

Johnson-Laird (1989) considers the creation of the mental model as a structure for comprehension and inference in general, for all agents who use language, but does not preclude its application in other domains. Whilst he particularly focused on classical syllogisms, he made clear that his theory was intended as a general explanation of thought (Manktelow and Jones 1987). Johnson-Laird's focus was on the interaction between any two agents via language. This differs from the following theorists who focus on a user interacting with a device or system.

2.1.1.2 Bainbridge (1992)

Focusing on operators of complex systems, with a view to understanding cognitive skill, Bainbridge (1992) considered the way the term 'mental model' is used in cognitive processing too general. She offered three distinct ways the term makes a contribution: (1) knowledge base – described as knowledge of the permanent or potential characteristics of some part of the external world; (2) working storage – described as temporary inferred knowledge about the present or predicted state of the external world (equivalent to Johnson-Laird's (1983) notion of mental models); and (3) meta-knowledge – described as knowledge of the outcomes and properties of the user's own behaviour. Bainbridge (1992) describes the knowledge base as equivalent to long-term memory and proposed this is responsible for inference and prediction. The working storage is interpreted as a knowledge structure of inferred knowledge, derived from the knowledge base in form that the user can work with in the current context. Bainbridge (1992) identifies that for cognitive skill, knowledge bases should be sufficiently developed to infer states and anticipate events, and meta-knowledge should be fully developed.

Bainbridge (1992) proposes 'processing modules' as the mechanism that meets particular cognitive goals. She hypothesized that multiple processing modules work simultaneously at different levels, responding to different strategies chosen by the operator to fulfil particular goals. The information needed is actively sought out, either in existing knowledge bases, the environment, or through further processing. The modules communicate with each other via working storage, with answers to lower level goals becoming input to fulfil higher level goals.

Like Johnson-Laird (1983), Bainbridge (1992) emphasizes the recursive nature of the process, and considers reference to background knowledge as key. In addition, she highlights information in the world as an important information source.

Bainbridge's (1992) focus, unlike Johnson-Laird (1983), addressed how operators seek out answers to specific goals. The emphasis on operator strategies may be specific to users of complex systems, as evident in Moray's (1990) theory.

2.1.1.3 Moray (1990)

Moray (1990) used lattice theory to represent mental models held by an operator of complex systems. He presented an elegant approach to cognitive processing, based on operators selecting a strategy associated with one of Aristotle's four causal hypotheses. Moray (1990) provides the following example to distinguish between causes:

> …when considering a switch causing a pump to operate, he (the operator) may consider; A *formal* cause (because it is in the 'on' position), a *material* cause (because it

closes a pair of contacts), an *efficient* cause (because it allows current to flow through the pump), or a *final* cause (because cooling is required). (p. 579)

Moray (1990) suggested that a mental cause lattice (MCL) existed for each cause, resulting in a formal cause lattice (FCL), material cause lattice (MaCL), efficient cause lattice (ECL) and a final cause lattice (FiCL). Each lattice was considered a hierarchy, with the high level goal at the top and extremely detailed information at the lowest level.

The chosen strategy activates cognitive processing within a specific MCL, allowing the operator to travel up and down the hierarchy in the lattice, and between lattices where one MCL links to an alternate MCL. The latter case represents the operator switching strategies to find a particular bit of information.

Moray (1990) proposed each MCL is derived from a shared 'physical system' lattice (PSL). This is based on physical relations between parts of the system based on experience or system specifications. He proposed that the relationship between these groups, and from each group to the PSL, is via 'holomorphic mappings', which play a significant part in the explanation and reduction of errors. For example, inaccuracies in the PSL would be carried into the MCLs. This one-way relationship differs from the recursive process proposed by Johnson-Laird (1983) and Bainbridge (1992).

It is proposed by the author that each of the four MCLs is equivalent to internal knowledge structures which allow inference and prediction in the sense of Johnson-Laird (1983), and can be used as 'working storage' as identified by Bainbridge (1992). These MCLs also reference some sort of 'background' knowledge structure in the form of the PSL.

Moray (1990) focused on troubleshooting problems in complex systems and was concerned with how 'answers' are found. Like Bainbridge (1992), the importance of goals and strategies is marked.

2.1.1.4 Summary of Comparison of Theories of Cognitive Processing

It is clear from the preceding discussion, that what can be termed 'background knowledge' in the form of 'knowledge base, PSL' or 'world model', plays a vital role in cognitive processing, both as an information source and reference for inferred knowledge structures such as 'mental models', 'working storage' or 'MCLs'. Bainbridge (1992) requires background knowledge to be 'sufficient', whilst Moray (1990) stresses the importance of its accuracy.

Figure 2.1 provides a schematic of the three theories discussed, to visually emphasize the similarities and differences. The area within the solid circle includes the processes and types of knowledge structures common to all people. The broken ring surrounding this represents people's 'background', which will vary based on experience and includes the various types of 'background knowledge'. This is termed the 'background bias ring', in this chapter, as differences in its content will 'bias' the inferred knowledge structure.

We can see from Figure 2.1 that Johnson-Laird (1983) and Bainbridge (1992) emphasize a recursive nature to the relationship between background knowledge and inferred knowledge structures that support inference and prediction, but also include additional knowledge structures which play different roles. Moray (1990) describes

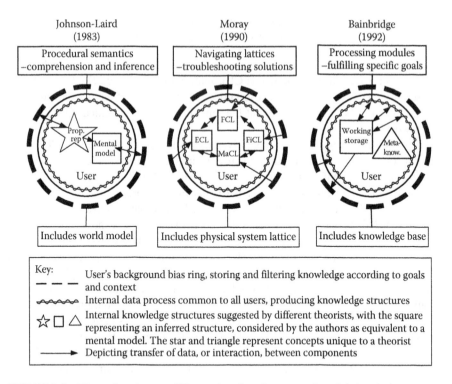

FIGURE 2.1 Illustrating the way different theorists place mental models in relation to other knowledge structures.

a one-way relationship and does not present additional knowledge structures, but instead four 'inferred' knowledge structures based on different causal strategies between which the operator can switch.

The different functional focus by different theorists suggests that different methods of processing may come into play in different contexts, meaning the theories discussed may all be relevant to an 'umbrella' theory related to multiple contexts. Johnson-Laird's (1983) focus on linguistics highlights the importance of communication between agents, which has relevance to any domain where linguistic communication occurs (both verbally and textually). Moray's (1990) focus on operators holding an accurate model of the relationships found in physical systems is relevant for interaction between humans and devices. Bainbridge broadens the concept by relating the background knowledge to any relevant elements in the 'external world'.

Whichever theorist is most applicable in a particular domain or context, it is clear that background knowledge is important in the construction of inferred knowledge structures which allow inference (what is termed in this chapter as 'mental models'). As such, the significance of background knowledge and how it interacts with mental models warrants further understanding, if the notion of mental models is to be applied pragmatically.

2.2   IMPORTANCE OF ACCURACY IN MENTAL MODEL DESCRIPTIONS: THE DEVELOPMENT OF AN ADAPTABLE FRAMEWORK

In this section, we develop a framework from the schematic shown in Figure 2.1, which identified the influence of 'background knowledge' on the content of inferred knowledge structures. Variations in background knowledge, therefore, bias the mental model constructed. It is suggested the same affect occurs when an analyst attempts to access a mental model, with their 'background knowledge' influencing the way data are interpreted. This view follows from that stated by the argument put forward by Wilson and Rutherford (1989) that an analyst's intention to access a user's mental model can result in the erroneous conclusion that their description is evidence of its capture. The following framework intends to emphasize the difficulties in accurately capturing (or constructing) a mental model, due to inherent biases in the methods for communicating mental models. The first part of this section will briefly describe cognitive bias and its existing role in mental model research. The meaning in this chapter of the terms 'filtering' and 'bias in interpretation' follow as the key concepts in the proposed framework. The second section will use the case study of Kempton (1986) to illustrate the layers of bias inherent in methodologies chosen to access internal constructs. The third part will consider the perspectives from which mental models are considered, illustrating the intended points by use of the framework.

2.2.1   BIAS AND FILTERING WHEN CONSTRUCTING OR ACCESSING MENTAL MODELS

Bias can be defined as 'An inclination toward a position or conclusion; a prejudice' (Reber 1985). Tversky and Kahneman introduced the notion of cognitive bias in 1974 in their seminal chapter which emphasized how heuristics are employed when people need to make judgements under uncertainty. They highlighted (1) representativeness, (2) availability of scenarios and (3) adjustment from an anchor as three highly economical 'rule of thumb' heuristics that are usually effective, but lead to systematic and predictable errors.

A major emphasis in the subject of 'bias' in the mental model literature has been on 'belief bias', the effect of believability of an outcome on whether it is accepted as true (e.g. Oakhill et al. 1989, Santamaria et al. 1996, Klauer et al. 2000, Quayle and Ball 2000, Klauer and Musch 2005). Other biases that have received attention include negative conclusion bias (Evans et al. 1995), matching bias, the effects of temporal order and suppositional bias (Ormerod et al. 1993), bias in meta-propositional reasoning (Schroyens et al. 1999), social, contextual and group biases (Jones and Roelofsma 2000). With the exception of Jones and Roelofsma (2000), who consider 'shared mental models' in teams, the goal of the literature has been to test mental model theory derived from Johnson-Laird (1983) on its ability to explain bias effects found in experimental data, with a view to amend, extend or abandon with alternate theories (Oakhill and Johnson-Laird 1985, Oakhill et al. 1989, Newstead et al. 1992, Newstead and Evans 1993, Ormerod et al. 1993, Evans et al. 1995,

Santamaria et al. 1996, Klauer et al. 2000, Quayle and Ball 2000, Klauer and Musch 2005, Schroyens et al. 1999).

Kahneman and Tversky (1982:201) class the deliberate manipulation of mental models as an important and distinct 'simulation heuristic' used particularly in: (1) prediction, (2) assessing the probability of a specified event, (3) assenting conditioned probabilities, (4) counterfactual assessments, (5) assessments of causality. Whilst emphasizing how other types of heuristics may interact with and affect the construction of mental models, they do not consider it necessary that mental simulation theory incorporate an explanation of these effects (Kahneman and Tversky 1982). In this chapter, the same sentiment is adopted and recognition of the 'risk of bias' to mental models research in general, rather than the explanation of specific bias types within a particular mental model theory, is the main concern.

To illustrate the importance of considering bias in mental model research, this chapter focuses on three broad categories: (1) the result of experience when interacting with the world (background bias), (2) that which comes into play when interacting simultaneously with another agent (social bias) and (3) when interacting separately, through some form of cognitive artefact (cognitive artefact bias). The first category is based on the conclusion from the previous section that differences in agents' backgrounds influence the nature of inferred knowledge structures (derived from Johnson-Laird 1983, Moray 1990, Bainbridge 1992). The second category builds upon Johnson-Laird's emphasis on linguistics and therefore communication. The latter extends from this to represent non-synchronous communication, which, by not allowing immediate interaction to check understanding, is considered to be particularly prone to bias in interpretation.

The framework offered will propose a set of 'tree-rings' surrounding each agent (i.e. the user, analyst, instructor, intelligent device, and so on), as depicted in Figure 2.2. These rings represent the agents' background (in terms of experience and knowledge), their social interactions (including communication and behaviour) and,

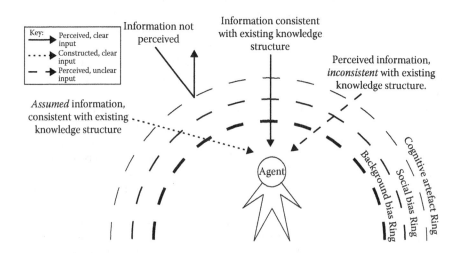

FIGURE 2.2 Depicting how information in the world can be filtered in different ways.

where relevant, their use of cognitive artefacts for communication or as an information source. The background ring is closest to the agent, and will always be present. The social and cognitive artefact rings may be present in various configurations depending on the type of interactions in which the agent is involved. The rings are represented as dashed lines to signify 'filters' allowing selected information to pass to the agent. The arrows portray different ways in which information from the world is treated by the combined effect of the filters. The key point to note is that information intake is not treated as the dichotomy of accepted or rejected, but according to its consistency with existing or alternative knowledge structures. The implication is that inconsistent information may not be rejected, but merely treated differently to consistent information. Assumed (rather than perceived) information, consistent with the existing knowledge structure, is also reflected as an input for the construction of functioning mental models. Different treatments of information are likely to affect the clarity of the input when forming knowledge structures (represented by different arrow styles in Figure 2.2).

As shown in the previous section in the theories of Bainbridge (1992) and Moray (1990), the idea of goals and strategies plays a significant role in resulting inferred models. In different contexts, or when the agent has different aims, the ways information should be interpreted to form a functioning mental model, will therefore differ. This implies that the filters vary accordingly. It is suggested that this difference represents the combined influence of cognitive biases relevant to that context, and provides a generic term to account for these variation as 'bias in interpretation' (henceforth termed 'bias').

Consider the analogy of a bowl with a ball of putty pressed off-centre, causing an imbalance or 'bias'. The bowl can be considered the background or social bias ring, with the putty representing the agent's bias, formed by his or her goal in a particular context. As the size and position of the putty in the bowl determines the angle of imbalance, so too the amount and kind of background and social experience determine the strength and direction of the agent's bias.

Figure 2.3a shows a bowl with no bias, whereby thought, action or behaviour are steered equally in all directions. Figure 2.3b shows the effect of strong bias, steering thought, behaviour and perception in a single direction. Since the focus of this review

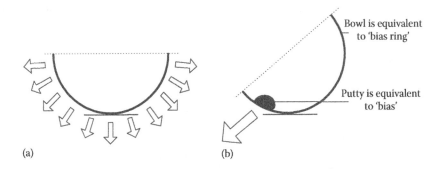

FIGURE 2.3 Providing an analogy for 'Bias rings'. (a) No bias in thought, behaviour and perception. (b) Biased thought, behaviour and perception.

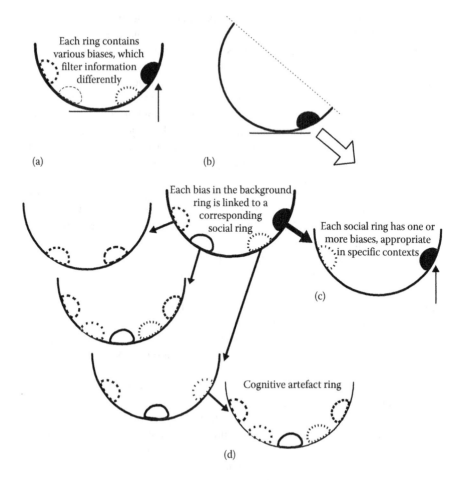

FIGURE 2.4 Proposing the interaction between bias rings and filters. (a) Selection of a particular background bias depends on the agent's goals and the perceived context. (b) Thought, behaviour and perception are the result of biased background and social filters. (c) Selection of a particular social bias depends on the selected background bias and the perceived context. (d) The format, structure and contents of any cognitive artefact produced are the result of filtering from the agent's background and social bias rings.

is on mental models, the bias discussed will be related to 'thought'. It should be made clear, however, that the interaction between perception, thought and behaviour is considered implicit.

It is assumed that each ring contains various biases, differing according to an agent's experience. Selection of a particular 'filter' is based on the agent's particular goal within the perceived context (see Figure 2.4a). A sole background ring exists for each agent, containing a set of biases that connect to corresponding social rings. The choice of social bias is similarly based on context and aims, filtering thought and behaviour from the agent, as well as the perception of information from the world (Figure 2.4b and c). The cognitive artefact ring is a special case of 'pre-filtered' information, arising from the background and social rings of the creator, and extends the chain in Figure 2.4d.

The aim of this section is not to evaluate in depth mechanisms to explain the bias in specific types of reasoning, but rather offer a high level mechanism that complements the idea of 'filtering' to emphasize the need to consider 'risk of bias' in mental models research. In each individual context, specific biases will apply and these will need to be understood within that context and chosen methodology. The framework developed is designed to depict the 'risk' of bias, as well as emphasize the perspective from which mental model structures are being considered, with a view to allowing commensurability in mental models research.

2.2.2 ACCURACY OF MENTAL MODEL CONTENT: A CASE STUDY OF KEMPTON (1986) ILLUSTRATING THE IMPACT OF METHODOLOGY

Kempton (1986) identified two different 'folk' theories of domestic heat control ('valve' and 'feedback'), interpreting data collected by in-depth interviews with users. Kempton (1986) suggested that a user's adopted folk theory for a device was the *driver* for their behaviour when interacting with the device. He argued that acknowledging this link could have practical applications in understanding and predicting behaviour that affects energy consumption. Having identified the considerable burden of domestic heating use on energy consumption, Kempton (1986) implied this could be reduced by ensuring users have the most energy-efficient folk theory of home heat control.

Kempton's (1986) study of home heating is chosen as an example to illustrate the impact of methodology on bias in mental models for three reasons. Firstly, as the area of study (energy-consuming behaviour) reflects the interest of this chapter; secondly, because he is clear in his description of methodology adopted and, finally, because the claims he makes about 'folk theory' are equivalent to the potential associated with the term 'mental model' as an inferred knowledge structure from which inferences and predictions can be made. Focusing on Kempton (1986), therefore, provides an opportunity to consider 'folk theory' in the context of mental model theory whilst developing the proposed framework.

Kempton (1986) paraphrased four findings of McCloskey (1983) regarding the folk theory of motion, as follows, to illustrate the basis on which he uses the term. The reference to institutionalized physics can ostensibly be substituted with the established 'expert theory' dependent on context.

> folk theory.... (1) is based on everyday experience, (2) varies among individuals, although important elements are shared, (3) is inconsistent with principles of institutionalized physics.... and (4) persists in the face of formal physics training. (p. 77)

Figure 2.5 depicts Kempton as the 'analyst' and represents two users as 'subjects', who have individual models of the operation of a home thermostat, but some key shared elements identified by the analyst. From this depiction, it is not possible to determine if the shared elements are part of a 'folk' theory, as only part (2) of McCloskey's (1983) four findings is established. A shared 'amateur' theory is suggested as an alternative that does not fulfil findings (3) and (4).

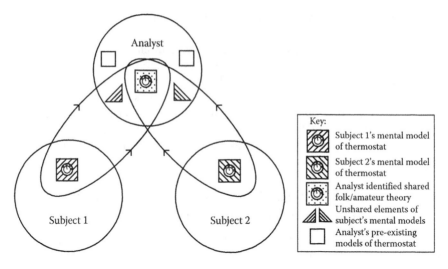

FIGURE 2.5 Depiction of the relationship between agents and models when identifying a shared theory of the operation of a home thermostat (based on Kempton's [1986] study).

In the depiction in Figure 2.5, the large circles represent the 'agents' involved. The squares within the circles represent the agents' internal 'mental model' and contain an image of the source device (in this case a thermostat dial). The interconnecting ovals represent that a process has taken place for one agent to access the model of another agent. The arrows flow from the source model (which has been accessed) to the resultant model (which is held by another agent and based on the accessed model). The key point to note is that each agent holds a different model.

For completeness, Figure 2.5 also shows the unshared elements of each subject's models, as well as pre-existing models that the analyst may have had (such as from interaction with an expert, or their own experience). In Kempton's (1986) study, the significant model is the 'shared theory' held by the analyst and formed by elements that were present in both subject's individual models. How this resultant model is constructed and the processes involved in accessing another agent's model will be discussed next.

2.2.2.1 Bias When Accessing Another Person's Mental Model

Kempton (1986) accessed his subject's mental models through in-depth interviews, where participants were asked to explain their understanding of how the home-heating thermostat operated. Particular focus was placed on the relationship between users' comfort and their interaction with the thermostat control. Evident in the subject's transcripts was the use of analogies to explain their understanding of the thermostat (such as a gas burner or water valve for 'valve theory'). This suggested that subjects were basing their individual models of thermostat control on previous experience. To infer two theories of thermostat control, it is reasonable to assume that Kempton's 'valve' and 'feedback' examples were similarly influenced by his existing knowledge and previous experience when interpreting users' statements.

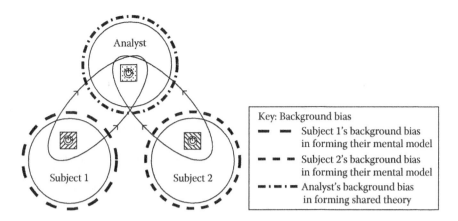

FIGURE 2.6 The influence of 'background bias' in forming an agent's mental model is shown as a patterned ring around each agent.

The impact of existing knowledge and experience in the construction and access of mental models has been reiterated in the literature (de Kleer and Brown 1983, Rouse and Morris 1986, Wilson and Rutherford 1989, Payne 1991, Bainbridge 1992, Smith-Jackson and Wogalter 2007, Zhang et al. 2010, Ifenthaler et al. 2011, Zhang and Xua 2011) and formed the conclusion of section 2.0, based on the work of Johnson-Laird (1983), Moray (1990) and Bainbridge (1992).

To represent the effect of the agent's background, in terms of experience and knowledge, on the construction of a mental model, a 'background bias' ring has been placed around each agent and depicted in Figure 2.6. The different ring patterns denote that each agent will have a background specific to them. As such, presented with the same device, different agent's interpretations of its workings will vary according to the variations in their background. The 'background bias' ring is in essence a filter, which comes into play when processing information from the world. This is similar to the concept of schema (Bartlett 1932, Stanton and Stammers 2008).

When interviewing subjects, Kempton (1986) noted that many showed insecurity in their descriptions, presumed by Kempton to be due to embarrassment about their incomplete knowledge. It is also evident from the transcript elements presented that Kempton's ability to adjust his questioning according to the subject's responses was key in clarifying and bringing out elements of the individual's model. Kempton also stated that he started the study with a larger number of exploratory interviews, but the lack of specific focus on home heat control meant they were inadequate for accessing an individual's specific device model. The effect of the social interaction between analyst and subject, and the skill and focus of questioning, cannot, therefore, be ignored and are considered to be examples of 'social bias'. This sentiment is similarly reflected in the literature (Rouse and Morris 1986, Wilson and Rutherford 1989, Payne 1991, Bainbridge 1992, Kennedy and McComb 2010, Cutica and Bucciarelli 2011, Frède et al. 2011) and is inferred by the work on linguistics by Johnson-Laird (1983).

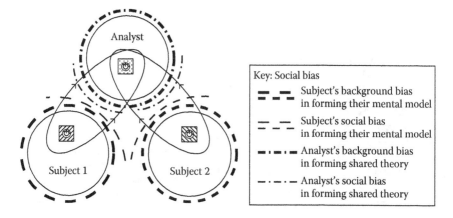

FIGURE 2.7   The influence of social bias on forming an agent's mental model.

Figure 2.7 represents the effect of social bias as an arc around the agent and between communicating agents. An arc, rather than a ring is used to highlight social bias, which only comes into play during an interaction with another agent. The arcs are patterned according to the agent, to show they are individual and specific to that agent, and related to the background bias ring.

Since the way an agent interacts may change according to the social context, and their assessment of what the other agent will understand or relate to, it is suggested that interaction with different agents will produce different 'social biases'. As was described in Section 3.1, it is proposed that these differences are ultimately influenced by the agents' background and experience. This means the 'background bias' ring not only has primary influence on the construction of mental models, but also affects the 'social bias' ring (as was shown in Figure 2.4). The 'social bias' ring essentially determines the ease with which one agent can access the mental model of another agent.

It has been proposed that the construction of mental models by an agent and their access by another agent is subject to biases relating to social/communication and background/experience. These influences have been represented schematically, in what shall be termed a 'tree-ring profile', in Figure 2.8. A tree-ring profile is constructed by aggregating the bias rings in order of interaction between agents, starting with the source model. Where more than one ring is shown, an interaction with another agent has taken place.

The mental models belonging to two different subjects vary in their construction according to the different bias of their unique background rings (see Figure 2.8). Kempton's shared theory is depicted as exposed to considerably more bias. As his theory is built up from more than one subject's mental model, the background bias surrounding the central model is made up of more than one 'pattern'. Similarly, the social bias of more than one subject is shown in the second-from-centre ring. The final two rings represent the influence of Kempton's own social bias (medium, line-dot pattern) when interacting with the subjects, and finally his own background bias

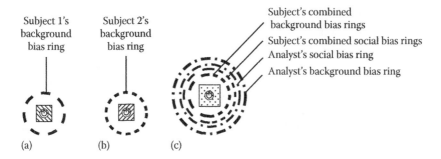

FIGURE 2.8   Tree-ring profiles showing the layers of bias that alter the construction and access of an agent's mental model. (a) Subject 1's mental model of thermostat depicted as a tree-ring profile. (b) Subject 2's mental model of thermostat depicted as a tree-ring profile. (c) Analyst identified shared folk/amateur theory of thermostat depicted as a tree-ring profile.

(thick, line-dot pattern) when constructing the shared theory from the common elements of the subject's individual mental models (see Figure 2.8).

We can see from the tree-ring profiles in Figure 2.8 that variations in models are inherent of variations in individuals. The process of accessing another individual's mental model, that is, the methodology of access, is subject to many layers of bias which by implication distort an analyst's description from a 'true' representation of an individual's mental model.

Whilst Kempton (1986) was not claiming to represent any one individual's mental model, but a 'shared theory', he nevertheless depended on accurate extraction of subject's individual mental models in order to identify common elements to form his hypotheses. The next section considers the perspectives from which analysts access mental models, positioning Kempton's individual and shared theory within these.

2.2.3   ACCURACY IN DEFINITION: THE PERSPECTIVE FROM WHICH DATA IS GATHERED

We have shown how the methodology undertaken in mental model research affects the risk of bias and therefore confidence in the accuracy of the knowledge structure. In the introduction, we emphasized how the literature has not reached a consensus on the definition of a mental model. This section argues the importance of characterizing mental models based on the perspective (in terms of model source and type) from which they are considered. To do this, existing definitions of mental models will be represented schematically to highlight the similarities, differences and ambiguities.

Norman (1983) provides a clear set of definitions relating to the concept of a mental model when interacting with devices. These will be contrasted with those provided by Wilson and Rutherford (1989) to show the variety and potential ambiguity in definition. These definitions have been chosen on the basis that they appear in articles with high citation ratings, implying widespread influence on subsequent research.

2.2.3.1　Norman (1983)

To prevent confusion when undertaking research or discussing mental models, Norman (1983) emphasizes the importance of distinguishing between the following four conceptualizations: (1) the *target system*, 't', which is the system that the person is learning or using; (2) the *conceptual model* of that target system, 'C(t)', which is invented by instructors, designers or scientists to (ideally) provide an accurate, consistent and complete representation of the target system; (3) the *user's mental model* of the target system, 'M(t)', which is 'the actual mental model a user might have', and can only be gauged by undertaking observations or experimentation with the user and (4) the *scientist's conceptualization* C(M(t)) of the *user's mental model*, which, by implication, is the mental model a user is *thought* to have.

These definitions have been represented schematically by the author in Figure 2.9. Note that M(t) is the only term where interaction with another agent is explicitly specified. It is not clear from Norman (1983) how C(t) or C(M(t)) is constructed. It might be presumed that there was interaction with the device itself, with other agents such as experts on either the user or the device, or with cognitive artefacts such as

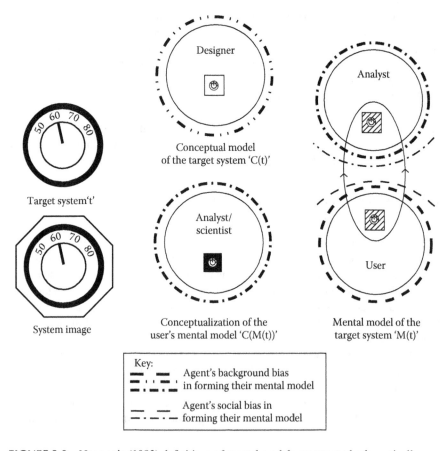

FIGURE 2.9　Norman's (1983) definitions of mental models represented schematically.

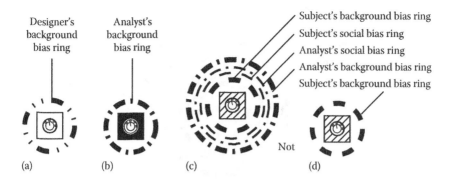

FIGURE 2.10 Tree-ring models representing the bias explicit in Norman's (1983) definitions. (a) Tree-ring profile for C(t). (b) Tree-ring profile for C(M(t)). (c) Tree-ring profile for M(t). (d) False tree-ring profile for M(t).

user guides, market research reports or technical specifications. Because the method for construction is not explicit, the influence of bias rings cannot be represented beyond the background of the analyst and designer (see Figure 2.10). This means that subsequent researchers using the same (quite well specified) terminology may do so with significant variation in the set of 'bias rings' impacting the access or construction of the mental model sought.

Figure 2.10 depicts the difference between a concept with a stated methodology, and that without, when looking at the two examples of M(t). Norman's (1983) definition provided the insight that M(t) can only be gauged by some methodology, rather than accessed directly. Since all of Norman's definitions apply to mental constructs, the presumed or recommended method of access would benefit others who wish to use the defined construct in a consistent way.

Comparing the tree-ring models in Figure 2.10 to those in Figure 2.8, we can see similarities. Norman's (1983) M(t) is similar in the pattern of bias rings to Kempton's (1986) 'analyst identified shared theory', with the exception in Norman, since interaction with only one 'user' is presumed, that a single line style is used to make up the rings of the agent holding the source model. Kempton's (1986) subjects' individual mental models, are equivalent to Norman's refuted example of M(t). Since Kempton (1986) made clear his methodology in terms of in-depth interview and analysis of transcripts, Norman's (1983) interpretation of a subject's individual model could be represented as the tree-ring model in Figure 2.11.

The addition of a bias ring representing a 'cognitive artefact' is shown as a thin ring in Figure 2.11, to indicate that a model is likely to be understood, or at least validated, as the result of analysis of the transcripts, rather than instantaneously during interview. The form of the cognitive artefact is a 'bias', since different interpretations may be possible based on its qualities (i.e. text-based as opposed to graphical) and structure (alluded to by Rasmussen and Rouse 1981, Smith-Jackson and Wogalter 2007).

The two analyst background rings show that two separate steps took place to access the model. The first, interviewing the subject whilst creating a cognitive

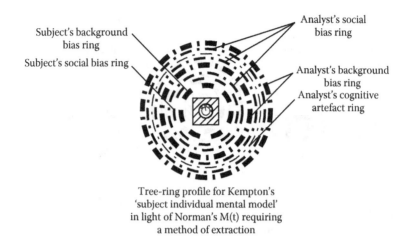

Tree-ring profile for Kempton's
'subject individual mental model'
in light of Norman's M(t) requiring
a method of extraction

FIGURE 2.11 Re-representation of Kempton's (1986) subject's individual mental model, in light of Norman's (1983) definition.

artefact for future analysis. The second, interpreting the cognitive artefact produced (see Figure 2.11).

2.2.3.2 Wilson and Rutherford (1989)

Wilson and Rutherford (1989) argue that psychologists and human factors analysts are talking about different things when discussing and researching mental models and that a unified terminology is needed if the cross-disciplinary efforts to research this construct are ever to be successfully applied or built upon. They offered the following definitions to this end:

> We use the term *designer's conceptual model* for the designer's representation of the user. The term *user's conceptual model* may be employed to mean the user's representations of the system, defined in terms as structured or loose as desired. We would reserve *user's mental model* to refer to descriptions of the user's internal representations which are informed by theories from Psychology. (p. 631)

The first two terms will be discussed initially and are represented both schematically and as tree-ring models in Figure 2.12. Since, both, the designer's conceptual model (DCM) and the user's conceptual model (UCM) are defined as 'representations', and no discussion of access is offered, the presumption is that these are internal representations that have not been gauged or shared by another agent. As such, the schematics in Figure 2.12 contain only one agent for each concept, and the tree-ring models show the effect of a lone background bias ring. It should be made clear that the DCM is not a model of a device (as with the UCM), but of another agent (in this case the user). To show this distinction, an image of a person, rather than a device, is shown in the square representing an internal model. It is assumed that Wilson and Rutherford (1989) consider the DCM akin to a stereotype of the user, based on an assumed user background. The application of this type of model could be in designing a device based on assumptions about what the user will need or understand. This is distinct

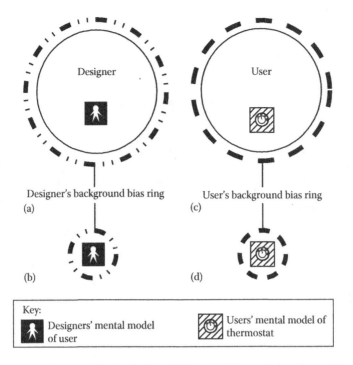

Designer's background bias ring

(a)

(b)

User's background bias ring

(c)

(d)

Key:

Designers' mental model of user

Users' mental model of thermostat

FIGURE 2.12 Wilson and Rutherford's (1989) definitions of mental model concepts, depicted as a schematic and tree-ring profile. (a) DCM. (b) Tree-ring profile for DCM. (c) UCM. (d) Tree-ring profile for UCM.

from the UCM which, if used as a design aid, could suggest how a user would *act* as a result of their internal model of a system. However, since both the UCM and DCM do not include a methodology for either 'access' or 'construction' in their definition, as with Norman's (1983) C(t) and C(M(t)), it would not be possible to determine the biases affecting their impact in any application. Wilson and Rutherford's (1989) UCM is equivalent to Kempton's (1986) subject's individual model and the false tree-ring profile of Norman's (1983) M(t) shown in Figure 2.10.

As Wilson and Rutherford's (1989) definition of a User's Mental Model (UMM) declared the existence of 'descriptions of the user's internal representations' (p. 631), it seems reasonable to assume that a cognitive artefact is included in this concept. Since these descriptions are 'informed by theories from Psychology' (p. 631), the involvement of an analyst with a background in psychology seems logical. What is not clear, from this definition, is if any other agents, or the target device, are involved in the UMM's construction, or if another agent is intended to access the resulting cognitive artefact.

Figure 2.13 shows four different interpretations of Wilson and Rutherford's (1989) definition of a UMM represented in schematics, with associated tree-ring profiles, as shown in Figure 2.14. Interaction with a device has not been represented, the addition of which could produce another four schematics based on those depicted.

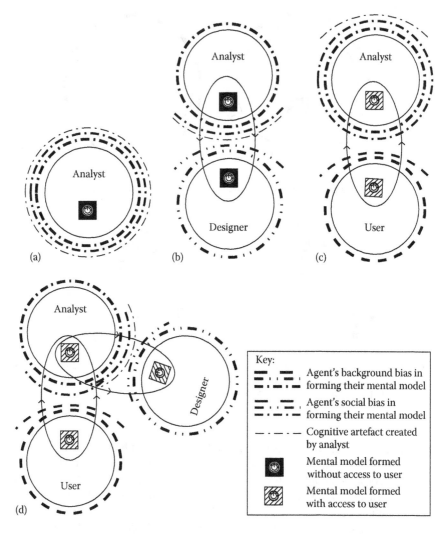

FIGURE 2.13 Four different schematics for Wilson and Rutherford's (1989) definition of a 'user's mental model'. (a) UMM1 – assuming no other agent involved. (b) UMM2 – assuming communication to a designer. (c) UMM3 – assuming user is model source. (d) UMM4 – assuming contact with user and designer.

UMM1 assumes that no other agent is involved, and is equivalent to Norman's C(M(t)), with the addition of a cognitive artefact bias ring. UMM2 builds on UMM1, but with the assumption that the descriptions produced are intended for access by a designer (see Figures 2.13 and 2.14). This would be a relevant example of conveying a mental model for practical application in device design, as recommended by Norman (1983) for the construction of a system image.

UMM3 assumes the analyst has derived a mental model after initial contact with a user. This is equivalent to Norman's M(t), with the addition of social bias and

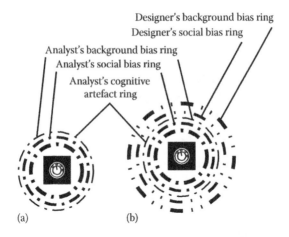

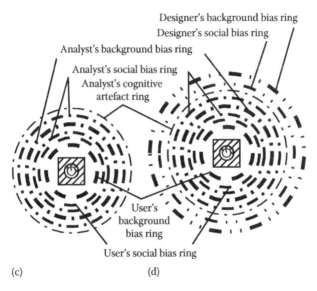

FIGURE 2.14 Tree-ring models representing the four different schematics for Wilson and Rutherford's (1989) definition of a 'user's mental model', highlighting the variation in biases based on interpretation. (a) Tree-ring profile for UMM1. (b) Tree-ring profile for UMM2. (c) Tree-ring profile for UMM3. (d) Tree-ring profile for UMM4.

cognitive artefact bias rings. Given Kempton's (1986) examples of 'shared theories' have been presented as a description, it is appropriate to add these rings, showing a similar tree-ring pattern to UMM3 (see Figure 2.15). The key distinction is the composite line styles in the central two rings representing the multiple users who held the model source for the shared theory.

UMM4 shows an extension of UMM3, whereby an analyst's model, which has been derived through contact with the user, is described in a cognitive artefact, which in turn is accessed by a designer (see Figures 2.13 and 2.14). Whilst it is

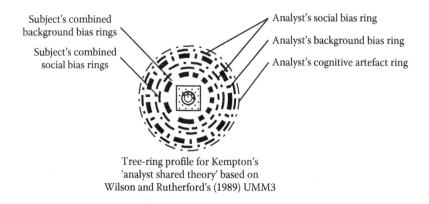

Subject's combined background bias rings

Subject's combined social bias rings

Analyst's social bias ring

Analyst's background bias ring

Analyst's cognitive artefact ring

Tree-ring profile for Kempton's
'analyst shared theory' based on
Wilson and Rutherford's (1989) UMM3

FIGURE 2.15   Re-representation of Kempton's (1986) analyst's shared theory, in light of Wilson and Rutherford's (1989) User Mental Model definition (assuming contact with the user).

evident that these assumptions cannot be attributed to the wording in Wilson and Rutherford's (1989) definition of the user mental model, this is included as a likely application in human factors research of this definition. As such, the differences clear in the tree-ring profiles shown in Figure 2.14 highlight graphically how the same definition can be subject to different interpretations by different researchers, preventing commensurability of findings due to the considerable differences in the layers of bias.

2.2.3.3   Summary of Comparison of Perspectives of Mental Models

By comparing the definitions of concepts provided by Norman (1983) and Wilson and Rutherford (1989), similarities in the concept of M(t) and UCM are shown, both signifying the internal representation on the user.

A key distinction is found between Norman's (1983) and Wilson and Rutherford's (1989) consideration of the model held by a designer. Norman offers the concept C(t) representing the target system, whereas Wilson and Rutherford describe the DCM, which represents the user as an agent, rather than an interpretation of the model a user may hold.

Despite the worthy intention of clarity in definition, Norman's (1983) concepts of C(t) and C(M(t)), and Wilson and Rutherford's (1989) concepts of DCM and UCM are rather vague in terms of their construction and possibilities for access. This prevents a precise understanding of the layers of interpretation to which any model description may have been subject. Whilst this vagueness may be considered appropriate, given that the intention of their formation was to provide tools for theoretical discussion, adopting these same terms when moving from theory to practice necessitates focus on specifics.

When depicting Wilson and Rutherford's (1989) UMM, the ambiguity the definition offered in terms of the original source for the model and the resulting recipient was obvious. The tree-ring profiles in Figure 2.14 show how the choice of source and recipient is crucial in understanding the layers of bias present. We have also shown how the tree-rings for Kempton's (1986) individual mental models and 'shared theory' can be positioned and extended, in line with these perspectives.

2.3 APPLICATION OF ADAPTABLE FRAMEWORK: CHARACTERING MENTAL MODELS BY PERSPECTIVE AND EVALUATING 'RISK OF BIAS'

A framework has been developed which has been useful in illustrating the differences between definitions of mental models, the perspectives from which data are gathered and the risk of bias in the accuracy of the knowledge structure description. This framework can go further than illustration and pragmatically aid commensurability and application.

To demonstrate this, schematics have been created to compare Kempton's (1986) 'shared theory' of thermostat function with Payne's (1991) 'mental model' of bank machine function (see Figure 2.16). Table 2.2 is populated from these schematics and details found in the respective theorist's papers. It has been divided into two main sections, which reflects the focus of this chapter. The first considers the perspective from which data are gathered, and that this is key for commensurability and application. Echoing the literature to date (e.g. Bainbridge 1992), it is clear that the context in which data are gathered on mental models is critical to applicability, since they represent a 'specific state of affairs' (Brewer's (1987) c.f. Stanton and Young (2000)).

The richness of the mental model description also constrains the applicability by others to design, behaviour change or instruction. We have seen that mental model definitions in the literature vary widely in meanings; Richardson and Ball (2009) have consolidated examples of mental representation terminology, emphasizing the range and subtle distinction between terms. It is proposed that for mental model research, where the same terms are employed with quite different meanings, it may be more prudent to specify the model source and recipient as an aid to commensurability than address consensus in terminology. The mental model source and recipient form a key point of distinction between definitions, but are often difficult to identify in the literature. The schematics allow model and source to be clearly identified.

The schematics also identify the number and character of intermediaries, and key aspects of their means of interaction with the source and recipient (single, multiple, synchronously or via a cognitive artefact). The risk of bias associated with these interactions is considered in the second section of Table 2.2. By using the schematics directly, or through the development of tree-ring profiles, the layers of 'risk of bias' can be considered in turn. The table's left-hand column prompts for specific information thought relevant to bias. The prompts are illustrative and are proposed as necessary rather than sufficient. They hold the role of directing consideration of relevant types of cognitive bias which may come into play during construction or access of the specific mental model. Examples of applicable bias are provided in the background ring of the source, and the social rings of both source and analyst.

The benefit of this framework is to provide a systematic means of identifying the risk of bias that is flexible enough to be applied to all types of mental model research. By doing so, efforts to mitigate or control for bias can be made, providing greater

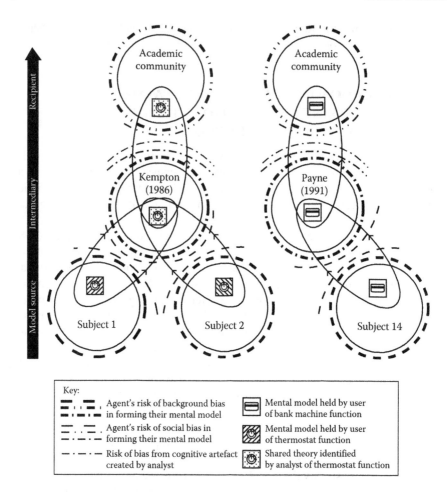

FIGURE 2.16 Comparing Kempton's (1986) 'shared theory' of thermostat function with Payne's (1991) 'mental model' of bank machine function, showing model source (subjects), intermediary (analyst) and recipient (academic community).

confidence that the accuracy of the mental model description reflects the construct held by the model source. Greater accuracy would be a considerable aid to applicability in design, behaviour change or instruction. Similarly, awareness of the limitations of accuracy allows the practitioner to adjust their expectations of the benefit of mental model descriptions to the prediction of human behaviour.

2.4 CONCLUSIONS

This chapter stated that the problem of commensurability in mental model research is based on a fundamental lack of consensus in its definition. It further argued that even with a consensus, existing approaches to definition would not be sufficient as

TABLE 2.2

Application of Proposed Framework to Kempton (1986) and Payne (1991) to Better Specify Mental Models for Commensurability, Considering the Risk of Bias in Interpretation and the Perspective from Which Data Are Gathered

Perspective of Gathering Data	Kempton (1986)	Payne (1991)
Mental model description	Valve	Distributed model
Context (domain, behaviour/task, goal)	• Domestic, patterns of adjusting temperature dial • Comfortable body temperature in the home at a reasonable cost	• The workings of high street bank machines • Operation to withdraw cash • Role of card, what happens during the transaction, how the machine operates
Mental model definition used (generic mental model, folk theory/conceptual model user mental model, etc.)	Shared theory	Mental model (individuals' specific knowledge)
Source (e.g. device user(s), designer, analyst)	Multiple users of domestic thermostats	Single user of bank machine
Intermediaries (e.g. analyst(s), system image, instructor)	Analyst	Analyst
Recipient (e.g. analyst, designer, academic community)	Academic community	Academic community
Layers of 'risk of bias'		
Source background ring(s) (e.g. number, defining demographics, relevant experience *applicable bias*)	• Multiple • Michigan homeowners (American) • Male and female *Representativeness* *Availability of scenarios* *Adjustment from anchor*	• Single • Lancaster Univ student (English) • No formal computing instruction *Representativeness* *Availability of scenarios* *Adjustment from anchor*
Source social ring(s) (e.g. number, method of communication, applicable bias, incentive) *applicable bias*	• Multiple, • Answers to in-depth interview *Confirmation bias* *Belief bias* *Consistency bias*	• Single • Answers to structured interview *Confirmation bias* *Belief bias* *Consistency bias*
Analyst social ring(s) (e.g. number, applicable bias, access method *applicable bias*)	• Multiple • In-depth interview *Order bias* *Belief bias* *Experimenter bias*	• Single • Semi-structured interview *Order bias* *Belief bias* *Experimenter bias*

(Continued)

TABLE 2.2 (*Continued*)

Application of Proposed Framework to Kempton (1986) and Payne (1991) to Better Specify Mental Models for Commensurability, Considering the Risk of Bias in Interpretation and the Perspective from Which Data Are Gathered

Perspective of Gathering Data	Kempton (1986)	Payne (1991)
Analyst background rings (e.g. number, defining demographics, relevant experiences, analysis method *applicable bias*)	• Single analyst • Anthropology • Energy and environmental studies • American • Infer folk theory from in-depth interview data using methods from Lakoff and Johnson	• Single analyst • Psychology, computing • English • Content analysed answers based on descriptions of devices, evidence of spontaneous construction, underlying analogies
Analyst social ring (e.g. number, *applicable bias*)	• Organization of textual information for *Cognitive Science* journal paper reviewers	• Organization of diagrammatic information for *Behaviour & Information Technology* journal paper reviewers
Analyst cognitive artefact ring (e.g. type(s), *applicable bias*	One-word analogous textual description	Basic diagram, key components and conduits for data transfer showing relationship between components. Data types held by each component specified
Academic community social ring (e.g. type(s), *applicable bias*)	Typical organization of information for *Cognitive Science* journal paper expected	Typical organization of information for *Behaviour & Information Technology* journal paper expected
Academic community background ring (e.g. type(s), *applicable bias*)	Cognitive science audience	Behaviour and information technology audience

the notion could not be applied pragmatically unless there was some confidence in the accuracy of the mental model accessed or constructed. The chapter highlights how the risk of bias to an accurate representation of a knowledge structure varies considerably, depending on the perspective used and the methodology of capture. Using a case study of Kempton (1986), an adaptable framework was developed to emphasize this graphically, and it was shown how the resulting schematic could be deconstructed in tabular form to specify more fully the similarities and differences between different research contributions.

The first part evaluated three major research figures in the field and their theories of the role of inferred knowledge structures in cognitive processing to conclude that variations in nature of 'background information' bias the content of inferred knowledge structures.

The second part of the chapter considered the role of cognitive bias in mental models research, with a view similar to Tversky and Khaneman (1974) adopted in this chapter, that cognitive bias interacts with mental simulation rather than necessarily being explained by existing mental model theory. The terms 'filtering information' and 'bias in interpretation' were described as precursors to development of an adaptable framework. Using the case study of Kempton (1986) and the textual definitions provided by Norman (1986) and Wilson and Rutherford (1989), the framework was developed to demonstrate the importance of methodology on the risk of bias, ultimately determining confidence in the construct description. Clarity in the mental model source and recipient for commensurability was also highlighted.

Part three brought together the findings of parts one and two to propose the criteria necessary to better specify mental models research with a view to commensurability and better confidence in levels of accuracy. The framework was used to derive the 'layers' of potential bias, providing a prompt for mitigation or appropriate caution when applying the mental model description.

The limitations to the work are as follows. A more extensive breakdown of the elements considered part of the 'background' ring and specification of the criteria which promote bias in the social and cognitive artefact rings is needed. In addition, a catalogue of types of cognitive bias specifically relevant to the methodologies typically adopted in mental models research, and means for mitigation would provide considerable benefit in improving the accuracy of knowledge structure descriptions.

Other areas central to mental models research such as memory, external representations and interaction with devices is either lightly touched upon or absent. The focus has also been on the mental models held by individuals rather than shared or team mental models. It is reasonable to be confident that the framework can accommodate these areas, but it is beyond the space limitations of this chapter, and provide an opportunity for further work.

The first task towards confidence in applying mental model descriptions for the benefit of design, behaviour change or instruction is an appreciation of the risk of 'bias in interpretation'. This chapter offers a systematic way of approaching this issue and the authors believe that through their framework, a different approach to specifying mental model research could further the goal to achieve commensurability.

The next step following the development of the tree-ring method, to further the objectives of this chapter, is to collect evidence of mental models of domestic heating systems in a real-world environment, with explicit awareness of the biases involved. Mitigation of biases can then take place where possible. Where this is not possible, the limitations of findings subjected to bias can be acknowledged. Conversely, intentional biases in line with the objectives of this chapter can be ensured. Chapters 3 and 4 demonstrate how identification of risk of bias using the tree-ring method can inform method development for access of knowledge constructs. This method was applied to householders in the United Kingdom to provide insights on how differences in present-day mental models of heating systems can influence behaviour with controls.

3 The *Quick Association Check* (QuACk)

A Resource-Light, 'Bias Robust' Method for Exploring the Relationship between Mental Models and Behaviour Patterns with Home-Heating Systems

3.1 INTRODUCTION

Energy consumption due to home heating is a key contributor to climate change, making up 58% of UK domestic energy consumption (Department of Energy and Climate Change 2011). The United Kingdom has legislated to cut carbon emissions by 80% by 2050 (Climate Change Act 2008) and domestic energy use makes up 25% of UK consumption (Department of Energy and Climate Change 2013). Lutzenhiser and Bender (2008) report that variations in domestic energy use are due to the behavioural differences of householders. There is a growing body of evidence that the design of home-heating systems and the interactions afforded, are key factors in energy-wasting behaviour (Brown and Cole 2009, Combe et al. 2011, Shipworth et al. 2010, Peffer et al. 2011, 2013, Glad 2012, Revell et al. 2014, 2015, 2016). Kempton (1986) highlighted the importance of differences in householders' mental models of thermostat function as a driver for differences in home-heating behaviour patterns. He proposed that ambiguous device design had a role to play in allowing variations in mental models to persist. Revell and Stanton (2014, 2015, 2016) investigated the present-day relevance of Kempton's (1986) work, adopting the methodology presented in this chapter. Their research focused on making more explicit the links between user mental models, behaviour and design at both the control device and the heating system levels.

Over the last 30 years, the notion of mental models has been applied as a strategy to predict how users will behave when interacting with a device and understand the reasons for inappropriate interaction. These range from research into consumer devices (Norman 2002), computer interfaces (Carroll and Olson 1987, Williges 1987) and complex systems (Moray 1990), and is particularly prevalent in domains such as transport (Stanton and Young 2005, Weyman et al. 2005, Grote et al. 2010) and command and control (Rafferty et al. 2010). The term 'mental model' is often used in the literature to describe internal constructs that differ significantly in terms of content, function or perspective (Wilson and Rutherford 1989, Richardson and Ball 2009, Revell and Stanton 2012). The method presented in this chapter is developed to capture a householders' 'device model' (Kieras and Bovair 1984) of the functioning of the home-heating system. The construct captured after data collection using QuACk, fulfils the criteria of Norman's (1983) definition of a 'user mental model' (UMM), that is, 'the internal representation held by an individual user', which can only be understood by an analyst through some method of extraction. The UMM output of QuACk, therefore, reflects the analyst's description of a UMM of the way the home-heating system functions. Following data collection, the method of data categorization presented by QuACk identifies what Kempton (1986) defines as a 'shared theory'. This can be thought of as a descriptive term that groups individual UMMs of heating devices by common elements. This allows exploration of association of specific categories with predicted trends in user behaviour with heating controls. These definitions could be considered by some readers to relate well to the notion of schemas. The authors adopt Brewer's (1987 c.f. Stanton and Young (2000)) distinction that 'schemas are generic mental structures underlying knowledge and skill, whilst mental models are inferred representations of a specific state of affairs'.

Kempton (1986) suggested that some shared theories held by users direct inappropriate energy use. Understanding why a user may behave inappropriately with their home-heating devices allows the possibility of mitigation. Understanding why groups of users display patterns of inappropriate home-heating behaviour allows mitigation strategies to be far-reaching. The types of strategies that could be adopted include instruction (Weyman et al. 2005, Sauer et al. 2009), automation (Stanton et al. 1997, Sauer et al. 2004, Lenior et al. 2006, Baxter et al. 2007, Larsson 2012), simplification (Papakostopoulos and Marmaras 2012), or various approaches of device and interface design (Williges 1987, Weyman et al. 2005, Baxter et al. 2007, Sauer et al. 2009, Branaghan et al. 2011).

Key methodological issues in ergonomic science are the need for valid and reliable methods that are quick and easy to apply (Stanton et al. 2014). Whilst various mental model elicitation methods exist (e.g. verbal protocol analysis, cognitive workthrough, the decision ladder from cognitive work analysis), depending on their fit with the target domain, they can be time-consuming to administer and resource-intensive during use. QuACk falls into what Annet (2002) considers an analytic method, arguing that the validity of methods of this type are based on its relationship to underlying theory as achievement of intended or presumed measurement is difficult to establish. These analysis methods have been adapted to understand mental models, but are not grounded in mental model theory. Stanton (2014) states that the extent to which many ergonomic methods have a sound basis in academic theory is questionable. He argues, however, that whilst reliability and validity of ergonomic methods is hard to come by, the goal is nevertheless pursued.

To confidently compare UMMs with user behaviour, an accurate description is desired but, like all knowledge structures, difficult to validate (Rouse and Morris 1986,

Wilson and Rutherford 1989, Bainbridge 1992, Richardson and Ball 2009, Revell and Stanton 2012). Whilst Kanis (2004) highlights bias as an issue in qualitative research, Revell and Stanton (2012) argue that methods designed to capture knowledge structures need to take into account bias both in the access and interpretation. The authors regard consideration of bias a key way of demonstrating a relationship with underlying mental model theory. Without this consideration, little confidence in the accuracy of a UMM description is possible, yet existing methods adopted to capture mental models do not emphasize this aspect. Revell and Stanton (2012) demonstrated that risk of bias to an accurate representation of a knowledge structure varies considerably, depending on the perspective used and the methodology of capture. A method to access UMMs would improve the validity of the construct and the chance of discovering association with user behaviour, where bias was actively considered in its development.

Hignett and Wilson (2004) argue that the relationship between academia and practice needs to be improved by strengthening the connection between theory and methods in practice. Desnoyers (2004) believes there is a definite need for valid methods of gathering qualitative information if the practitioner is to solve 'actual', rather than 'theoretical', problems. QuACks bring mental model theory closer to practice by providing a domain-specific focus to data collection and analysis, and providing clear guidance and instructions to mitigate bias in application.

Hancock and Szalma (2004) emphasize the need to embrace and integrate qualitative methods in ergonomics research. Hancock et al. (2009) argue that ideographic case representations are increasingly relevant for the design of human–machine systems, particularly as advances in technology begin to focus on exploiting individual differences. According to Flyvbjerg (2011), meaningful insights (that would be missed from context-independent findings) can be gained from rich data gathered from detailed, real-life situations. In response, QuACk was developed using case studies and participant observation as methods that provide rich feedback. This feedback was used to drive iterative developments to the prototypes.

Quack has been developed with the intention of pursuing the goal of addressing key methodological issues in ergonomics science found lacking in existing methods for capturing mental models. The initial aim of this chapter is to illustrate where and how methodological issues have been addressed (including steps taken towards analytic validity, behaviour pattern reliability and intra-analyst repeatability). Secondly, this chapter sets out to demonstrate the benefit of iterative development through a case study approach, by demonstrating the resulting improvements. Finally, this chapter discusses the potential to apply the findings of QuACk to energy-reducing strategies. The flexibility and generalizability of the method are also considered.

3.2   METHODS USED FOR THE DEVELOPMENT
AND EVALUATION OF QuACk

Figure 3.1 shows the stages in the development process of Quack, which is broken into three parts: (1) prototype development, (2) qualitative iterations and (3) reliability testing. The starting point was a literature review seeking previous work related to

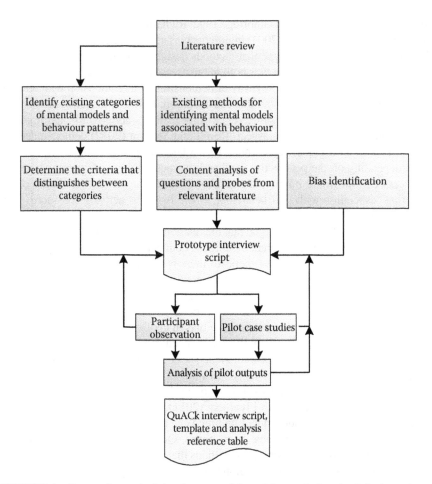

FIGURE 3.1   Process for method development of the quick association check for home heating (QuACk).

the association and categorization of users' models and behaviour, both, in general, and specifically with home-heating systems. Methods, questions and probes considered appropriate for the home-heating context, and shown to be successful in previous studies, contributed to the QuACk prototype. An iterative approach was adopted to refine QuACk, through qualitative methods such as participant observation and case studies. Finally, initial reliability tests were undertaken to instigate validation of the method. For brevity, each of the stages of development will be briefly described in turn, broken down according to the different cells in Figure 3.1. The relevant section headings have therefore been added to Figure 3.1 to aid navigation.

3.2.1   LITERATURE REVIEW

The human factors and cognitive science literature was reviewed (and is cited throughout) with three objectives: (1) to understand key considerations when conducting mental models research, (2) to identify existing methods for accessing mental models associated with behaviour and (3) to identify existing categories of mental models and behaviour patterns relating to the home-heating context. The conclusion of the first, as discussed in the introduction, was identification of the need to consider cognitive bias in data collection and analysis methods as a key variable that interacts with mental models of both the subject and the analyst, which aligns with the position put forward by Tversky and Kahneman (1974). Revell and Stanton (2012) describe how information communicated and received during any interaction is filtered by layers of background, social and cognitive artefact bias, distorting the resulting 'mental model description'. Any data collection method demands an interaction of some kind to take place. Subject and analyst bring their own experiences and expectations to any interaction, the communication of different aspects of knowledge are supported by the type of interaction adopted and the analysis possible is restricted or enabled by the form of data recorded and presentation of outputs. This work culminated in the development of the 'tree-ring method' for depicting bias in mental models research, which was adopted in the development of QuACk. Revell and Stanton (2012) provide a comprehensive debate and examples of applying the tree-ring method to different domains that the reader can refer to in more depth. The outcome of the second was used to determine the form of QuACk as well as to identify relevant questions and probes for content analysis (described in Section 3.2.2). The results of the third were used to form hypotheses for association and to understand the differences between existing categories of models and behaviour in the literature (see Section 3.2.3). These differences were used to inform data collection on appropriate variables and to aid development of an analysis table for distinguishing between categories (described in Section 3.2.6).

3.2.2   ASSESS METHODS FOR HOME-HEATING CONTEXT

The literature review failed to reveal a single method that enabled quick and easy exploration of association between mental models and behaviour. However, Rouse and Morris (1986) provided an excellent review of data collection methods for identifying mental models associated with actions, culminating in a succinct generic categorization of method types. Using this categorization, and seeking more recent work that fell within these groups, the authors assessed these methods' appropriateness for understanding user mental models of home heating and associated behaviour of home heating. The behaviour considered was a typical week in the naturalistic domestic setting, as this allowed direct comparison with Kempton's (1986) behaviour pattern examples. The authors considered the ease and time taken to collect data, the ease with which data could be analysed to explore association, and the costliness of the approach. The results of this evaluation are shown in Table 3.1.

From Table 3.1, the authors identified interviews and questionnaires as the most suitable method type to achieve the authors' goals, whilst recognizing a need to clarify and reduce the time for analysis. This format allowed flexibility to ground the data collection process in mental models theory. A further search of the literature to identify options for the format and structure of the questionnaire and the content

TABLE 3.1

Methods for Identifying Mental Models Associated with Behaviour by Rouse and Morris (1986), Evaluated for Domestic Home-Heating Context, Based on Speed, Ease and Cost of Data Collection and Analysis

Method Types Identified by Rouse and Morris (1986)	Ease of Data Collection for Home-Heating Context	Time Taken to Collect Data in Home-Heating Context	Ease of Data Analysis for Home-Heating Context	Estimated Cost in Home-Heating Context
Inferring characteristics via empirical study (Inferring model held by measuring related variable, in controlled experiment, e.g. Kessel and Wickens 1982, Mathieu et al. 2000, Langan-Fox et al. 2001, Sarter et al. 2007)	• Possible to remotely collect data on thermostat set points, boiler operation (as related variable). • Requires technical equipment and expertise. • Requires access to dwellings of sample.	• Between 1 week and 3 months (over winter) to collect 'typical' behaviour for a week. • However, multiple households could be collected simultaneously.	• Possible high volume of data, requires skilled processing to achieve a form ready for analysis. • Using behaviour as variable presupposes a link with mental models, so relationship cannot be explored by this variable alone.	• High cost of data collection equipment. • High cost of installation and technical support. • Cost associated with data processing and analysis (computer, software, data processing expertise).

(Continued)

TABLE 3.1 (Continued)

Methods for Identifying Mental Models Associated with Behaviour by Rouse and Morris (1986), Evaluated for Domestic Home-Heating Context, Based on Speed, Ease and Cost of Data Collection and Analysis

Method Types Identified by Rouse and Morris (1986)	Ease of Data Collection for Home-Heating Context	Time Taken to Collect Data in Home-Heating Context	Ease of Data Analysis for Home-Heating Context	Estimated Cost in Home-Heating Context
Empirical modelling (Algorithmically identifying the relation between users' perceptions and actions – only possible in simple scenarios, where user perceptions can be assumed, and resulting actions have not alternate explanations, e.g. Jagacinski and Miller 1978)	• Home-heating context is too complex, as users may vary in: (a) what they pay attention to, and (b) how they perceive (visual, haptic, thermal) before making adjustments. • Actions could be attributed to alternate explanations (e.g. multiple users, habit, and so on). • Specialist technical equipment and expertise required to capture attention, perception and actions.	• Between 1 week and 3 months (over winter) to collect 'typical' behaviour for a week. • However, multiple households could be collected simultaneously.	• High volume of data, requires skilled processing to achieve a form ready for analysis. • Using behaviour as variable presupposes a link with mental models, so relationship cannot be explored by this variable alone. • Complex analysis due to number of perception and attention and behaviour variables in naturalistic setting.	• High cost of data collection equipment. • High cost of installation and technical support. • Cost associated with data processing and analysis (computer, software, data processing expertise).

(Continued)

TABLE 3.1 (*Continued*)

Methods for Identifying Mental Models Associated with Behaviour by Rouse and Morris (1986), Evaluated for Domestic Home-Heating Context, Based on Speed, Ease and Cost of Data Collection and Analysis

Method Types Identified by Rouse and Morris (1986)	Ease of Data Collection for Home-Heating Context	Time Taken to Collect Data in Home-Heating Context	Ease of Data Analysis for Home-Heating Context	Estimated Cost in Home-Heating Context
Analytical modelling (Using theory/data to assume the form of different mental models, then comparing these 'model' forms to user performance, e.g. Anderson 1983, Yakushijin and Jacobs 2011)	• Possible to remotely collect data on thermostat set points, boiler operation (as related variable). • Requires technical equipment and expertise. • Requires access to dwellings of sample.	• Between 1 week and 3 months (over winter) to collect 'typical' behaviour for a week. • However, multiple households could be collected simultaneously.	• Possible high volume of user performance data, requires skilled processing to achieve a form ready for analysis. • Using behaviour as variable presupposes a link with mental models, so relationship cannot be explored by this variable alone. • Kempton (1986) provides example of form of thermostat behaviour and associated model, but parameters insufficiently specified for systematic categorization.	• High cost of data collection equipment. • High cost of installation and technical support. • Cost associated with data processing and analysis (computer, software, data processing expertise).

(Continued)

TABLE 3.1 (Continued)

Methods for Identifying Mental Models Associated with Behaviour by Rouse and Morris (1986), Evaluated for Domestic Home-Heating Context, Based on Speed, Ease and Cost of Data Collection and Analysis

Method Types Identified by Rouse and Morris (1986)	Ease of Data Collection for Home-Heating Context	Time Taken to Collect Data in Home-Heating Context	Ease of Data Analysis for Home-Heating Context	Estimated Cost in Home-Heating Context
Verbal protocol (Transcript of subject 'Thinking aloud' as they perform a task, e.g. Rasmussen and Jensen 1974, Norman 1983, Greene and Azevedo 2007, Ball and Christensen 2009)	 • Not appropriate for naturalistic home-heating context as home heat control task takes place in private dwelling, at potentially unscheduled/irregular times. • Subject needs to be trained in process of thinking aloud. • Analyst would need to be present in dwelling for duration, or subject trained to record protocols.	• Minimal 1 week per person if analyst present.	• Lengthy transcription times. • Analysis time may be lengthy, depending on criteria for categorizing models/behaviour. • Kempton (1986) provides example of form of thermostat behaviour and associated model, but parameters insufficiently specified for systematic categorization.	• Minimal (pen, paper, audio-recording device).

(Continued)

TABLE 3.1 (*Continued*)

Methods for Identifying Mental Models Associated with Behaviour by Rouse and Morris (1986), Evaluated for Domestic Home-Heating Context, Based on Speed, Ease and Cost of Data Collection and Analysis

Method Types Identified by Rouse and Morris (1986)	Ease of Data Collection for Home-Heating Context	Time Taken to Collect Data in Home-Heating Context	Ease of Data Analysis for Home-Heating Context	Estimated Cost in Home-Heating Context
	🙂	🙂	😐	🙂
Interviews/questionnaires (Analyst asks participants about what they think and how they behave, either verbal or written, e.g. Hutchins 1983, Kempton 1986, Stanton and Young 2005, Schoell and Binder 2009)	• Can be performed anywhere with minimal equipment. • Does not need to occur during task performance.	• Controlled by length of questionnaire/interview. • Typically less than 2 hours per subject.	• Lengthy transcription time, but can be influenced by design of interview. • Analysis time may be lengthy, depending on criteria for categorizing models/behaviour. • Kempton (1986) provides example of form of thermostat behaviour and associated model, but parameters insufficiently specified for systematic categorization.	• Minimal (pen, paper, audio-recording device).

of the questions and probes was then undertaken. This provided the data for content analysis described in Section 3.2.2.1.

3.2.2.1 Content Analysis of Previous Research

The work of Kempton (1986, 1987) and Payne (1991) both have high citation ratings and consider the mental models of everyday devices. As such, they were selected as a credible and representative sample of questions and probes for eliciting the mental models of domestic devices. Further, Kempton's (1986) work was focused on the domain of interest. The transcripts and questions provided in these texts were content analysed by question type, subject and inferred purpose for capturing mental model descriptions. These are summarized in Table 3.2 and informed the choice of questions and probes in the QuACk prototype, which used all question types, but was predominantly made up of open questions, direct questions and scenarios.

Additional considerations were made in the choice of probes. Firstly, Oppenheim (1992) emphasized how the data from hypothetical questions should be treated with care. To distinguish between typical, possible and hypothetical scenarios, the participants were asked how likely each scenario was for their lifestyle. This allowed the data to be assigned the appropriate validity, if more in-depth analysis of the transcripts was desired (see Revell and Stanton (2014), for an example of in-depth analysis of user mental models of home heating using data collected by QuACk). Secondly, although Kempton (1986) suggested analogies to his participants as an example that they could confirm or dispute, the authors felt this added a risk of bias that encouraged the participant to reframe their mental model in a form that was easy for the analyst to categorize. However, as Payne (1991) emphasized, analogies were frequently used by participants to describe their thinking; a direct question, rather than a leading question was used to elicit analogies, for instance, 'Can you think of any other device that works in the same way' rather than 'Does the thermostat work the same way as a gas stove'.

3.2.3 EXISTING CATEGORIES OF MENTAL MODELS AND BEHAVIOUR

To ground the method in mental models theory, existing categories of mental models and behaviour were sought. Kempton (1986) identified two key 'shared theories' of home heating, which he termed 'feedback' and 'valve'. Norman (1986) expanded these to include 'timer' and Peffer et al. (2011) refer to a 'on/off switch' model. Richardson and Ball (2009) provide useful descriptions of valve, timer and feedback theory (with the latter ambiguously labelled a 'switch theory'). These and Peffer et al.'s switch theory are summarized in Table 3.2, relating specifically to the thermostat as the key device to control home heating.

Kempton (1986) hypothesized the behaviour patterns associated with feedback and valve mental models of home heating. These patterns were based on thermostat set point controls over a week's period. Kempton (1986) proposed that users with a valve shared theory considered changes in the set point of their thermostat to be controlling the intensity of heat in their furnace (rather than maintaining a specific house temperature), so they felt the onus on them to ensure a comfortable home temperature. He anticipated valve behaviour patterns to show frequent, irregular

TABLE 3.2

Subject and Purpose of Probes Used by Kempton (1986) and Payne (1991)

Question Type	Topic of Enquiry	Purpose to Assist Data Collection for Mental Models
Direct question, e.g. 'How does the thermostat work?'	Device within a system	Variables which influence device function
	Device function	Model of how a device works
	impact of time	Cause and effect in mental model
	Dependencies	Cause and effect in mental model
	Experience with device class	Source of mental model
Open question, e.g. 'Why does the boiler turn off?'	All subjects	Reduce interview bias
Scenarios, e.g. 'It's a cold day and you want to warm up, what do you do?'	Typical	Likely behaviour, experience of consequences of behaviour
	Possible (speculation)	Reasonable speculation about behaviour, or consequences of behaviour (from which mental models could be inferred)
	Untypical/hypothetical (speculation, own behaviour, consequences of behaviour)	Speculation about behaviour or consequences of behaviour (from which mental models could be inferred)
	Varying conditions	Determining if mental model is influenced by specified variables
Leading questions, e.g. 'you mean the thermostat works like a gas stove?'	Suggesting analogies	Checking analyst's understanding of subject's mental model
	Suggest reasons for response (to be confirmed or otherwise by participant)	Check if variables of interest to analyst are significant, or facilitate the participant if having difficulty articulating response
	Providing options	Categorize response with regard to areas of interest to the analyst
Other techniques, e.g. 'Uh-huh'	Reassurance/encouragement	Reduce social bias
	Checking	Verify understanding
	Gauging importance	Verify significance of response to mental model held
	Paraphrase / summarize	Allow opportunity for participant to accept or correct interviewer's understanding

adjustments to the thermostat, and low settings overnight to save energy. Conversely, Kempton (1986) described that users with a feedback shared theory considered their responsibility merely to select the desired thermostat set point to ensure a constant house temperature, and the system would take care of the rest. He expected feedback behaviour patterns to show infrequent, regular adjustments to the thermostat, as changes only needed to be made in line with changes to occupants' activities. Kempton (1986) provided example outputs of thermostat set point changes over a week's period to illustrate the expected difference between users with valve and feedback mental models of their domestic heating system. Long-term behaviour patterns associated with switch and timer mental models were not evident from the literature, however. The four categories of mental model and two categories of associated behaviour patterns were used as a basis to: (1) infer behaviour patterns for the remaining categories; (2) identify the criteria that distinguish between categories. To help explore association between users' models of home heating and their associated behaviour, these patterns and criteria were used to aid development of probes and templates for data collection (see Section 3.2.5), as well as populate an analysis reference table for categorizing mental model and behaviour pattern outputs (see Section 3.2.6).

3.2.4 CONSIDER BIAS IN MENTAL MODELS RESEARCH

The literature review identified bias as a key issue when conducting mental models research and is a key distinguisher of QuACk as a method designed specifically for mental model capture, by focusing explicitly on bias to improve the validity of mental model output. The tree-ring method (Revell and Stanton 2012) was developed to consider bias when evaluating methods for capturing and analysing mental models. This method was applied to Kempton (1986) and Payne (1991), as their approaches for mental model access was relevant in terms of the source, intermediaries and recipient, to the authors' proposed form of QuACk. Revell and Stanton (2012) provide a detailed and comprehensive account of identification of the bias rings listed in Table 3.4, which relate to: (1) the background experience of both the analyst and participant; (2) the social expectations and means of communication of both the analyst and participant; (3) the structure of cognitive artefacts used in the interview and (4) the method of analysis of cognitive artefacts. These types of bias, and the strategy for mitigation or clarification, informed the choices made during the development of the QuACk prototype (see Table 3.4). In addition, the generic advice provided in Oppenheim (1992) on reducing bias in interview design was incorporated.

3.2.5 DEVELOPING DATA COLLECTION METHOD

Figure 3.1 shows how development of the data collection method was influenced by: (1) considerations of bias, (2) understanding of existing categorizations associated with home heating and (3) informed by content analysis of questions and probes of previous research. The components of the prototype are shown in Figure 3.2. The format of the prototype comprised a semi-structured interview, and involved two paper-based activities that required only pen, paper and post-it notes to undertake. A voice recorder was also used to record the participants' responses to allow more in-depth analysis if needed. This fulfilled the aim of QuACk to be a 'resource-light' method. The questionnaire prototype was just over 3 A4 pages, and was intended to take between 30 minutes to 1 hour to complete, including the paper-based activities, falling into Stanton et al.'s (2014) criteria of a method that you use when you don't have much time. Variations are expected due to variations in the complexity of the model held by the participant. The primary aim of QuACk was to allow exploration of association between user mental models of home-heating function and user

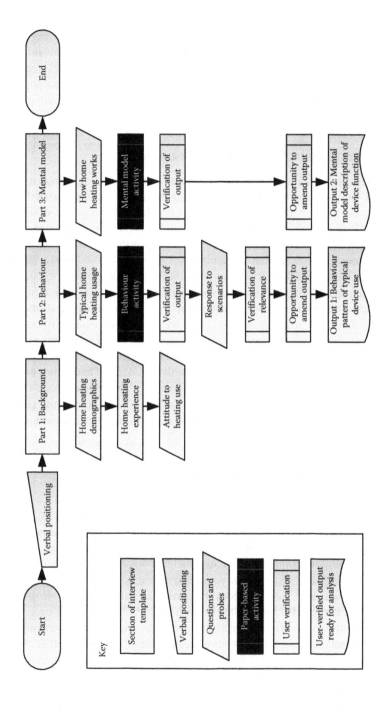

FIGURE 3.2 Components of QuACk prototype including elements informed by Payne (1991), Oppenheim (1992), Kempton (1986) and Revell and Stanton (2012).

behaviour with home-heating systems. Figure 3.2 shows that the key outputs of the method are a mental model description of device function and a behaviour pattern representing the self-report of home-heating use.

3.2.5.1 Paper-Based Activities

Two paper-based activities were included in the QuACk prototype. The first was a self-report of home-heating use, the second the development of a mental model description of the way the home-heating system functions. For the quick exploration of association, it is beneficial that the outputs are in a form ready for analysis. The activities were designed to produce the desired form of output in conjunction with the participant, allowing them to verify the content before the end of the interaction. This reduced the risk of bias in interpretation by the analyst as additional transformation of the outputs was not required. The structure of output 1 (behaviour pattern of typical device use) was designed to match the format produced by Kempton (1986) by showing changes in thermostat adjustment over a weekly period so anticipated patterns of use could be easily identified. The prototype initially required the week axis to be produced with pen and paper, but eventually evolved into a template axis which could easily be annotated (see Figure 3.3).

3.2.5.2 Verification of Outputs

From Figure 3.2, the emphasis on user verification is clear with five opportunities represented for the participant to make adjustments to the outputs. The authors believe this result in data that better reflect the participant's intention, than the expectations of the analyst. The emphasis on user verification is considered by the authors to be a distinct and necessary characteristic of QuACk that lacks prominence in other methods used to capture mental models. In addition to opportunities to amend responses, the QuACk prototype required the participants to deliberately assign a level of confidence to each information component of the outputs. This was added to help the analyst understand the credibility of any association (or lack of association) identified from analysis of the outputs alone. The process of output verification required the interviewer to paraphrase each element of the output in turn. The participant was asked to comment if this reflected what they imagined/thought (in which case a 'smiley' icon was added to that element), if they wished to amend what was represented

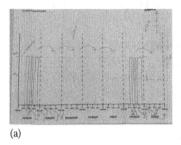
(a)

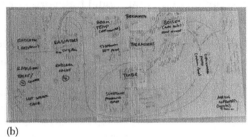
(b)

FIGURE 3.3 Example of QuACk outputs: (a) template with annotated self-report of home-heating use; (b) user mental model description of home-heating function.

(the amendment would then be made and a 'smiley' added), or if they felt they were uncertain if this reflected what they really thought (in which case a '?' was added next to the component). Examples of this annotation can be seen in Figure 3.3. The structure of output 2 (mental model description of device function) is generic and would suit contexts outside of the home-heating domain. The format was based on the mental model diagram by Payne (1991) for bank machines, the insight by Gentner and Gentner (1983) with regard to structure mapping and the emphasis by De Kleer and Brown (1983) on components and conduits in mental model descriptions. The output is similar to a concept map, but includes written rules and causal links. To produce this output, post-it notes, a pen and an A3 blank paper were needed (see Figure 3.3). By applying qualitative methods, the method evolved to comprise a participant information sheet, interviewer instructions, interview template, self-report activity template and mental model of home-heating function activity (provided in Appendix A). The iterations that led to this evolution are described in Table 3.6.

3.2.6 DEVELOPING ANALYSIS METHOD

A key feature of QuACk is a means to quickly explore association between mental models of device function and typical patterns of behaviour. Many data collection methods are open to multiple analysis types, without domain-specific guidance (e.g. verbal protocol analysis results can be categorized from multiple perspectives). Whilst this can be seen as a methodological strength for analysis by a general practitioner, the authors believe a domain-specific means of categorization allows 'quick' analysis in addition to mitigating for bias and facilitating commensurability. To achieve this, it was necessary for the analyst to easily distinguish between features to categorize the outputs. Table 3.3 was produced by referring to the criteria that distinguished the mental model descriptions provided by Kempton (1986), Norman (1986), Peffer et al. (2011) and Richardson and Ball (2009). The role of Table 3.3 is an analysis aid when examining the two verified outputs of QuACk following completion of an interview. It is divided into two sections corresponding to

---

TABLE 3.3
Description of Mental Model Categories of Home Heating from the Literature

Thermostat Theory	Description
Feedback	Turn the thermostat up and the heating stays on at full power until the temperature set is reached, then switches off (Richardson and Ball 2009)
Valve	Turn the thermostat all the way up and heat of a greater temperature is produced such that the room will get warmer faster (Richardson and Ball 2009)
Timer	Turn the thermostat up and the heating will stay on at the same temperature for a greater proportion of time (Richardson and Ball 2009)
Switch	Turn the thermostat sufficiently up, the heating will turn on. Turn the thermostat sufficiently down, the heating will turn off (inferred from Peffer et al. 2011)

these outputs: (1) the populated self-report template showing typical home-heating use, and (2) the mental model description of home-heating function.

The criteria identified comprised: (1) the thermostat set point, (2) variations in its value and pattern of adjustment, (3) the reason for its adjustment and (4) the role and function of related devices (e.g. boiler). For each mental model category of thermostat function (valve, feedback, switch and timer), Table 3.4 describes the form of the output that is expected. This is broken down by a set of criteria for each of the three outputs. Descriptions in italics represent compatible (rather than mandatory) responses that follow when a particular category of model is held. To explore association between users' mental models of home heating and their behaviour, the analysis table was constructed to guide the analyst in seeking out particular behaviour patterns and mental model elements and, where possible, categorizing them according to the shared theories from the literature (i.e. valve, feedback, switch, timer). The same category of shared theory found in both the self-report behaviour output, and the mental model description of home-heating function, would suggest an association may exist that warrants further investigation.

3.3   PILOT CASE STUDIES AND PARTICIPANT OBSERVATION FOR DATA COLLECTION

Good practice in methodological development is to undertake a pilot study to hone the content, structure and process Plant and Stanton (2016). To gain an insight on the flexibility of the QuACk prototype methodology, participants that varied in age, gender, house type and living circumstances were sought (detailed in Table 3.5) to encourage the development of a robust method that could cope with the variations inherent in the UK population. The target audience for participants was long-term UK residents with equivalent home heating setups (comprising gas central heating, combi-boiler, radiators, separate thermostat and programmer). The initial participant was selected based on availability to provide feedback on clarity, ease of answering questions, process of paper-based activities and the participant experience (in terms of the social interaction and feelings). Casual discussions with the target audience group regarding their behaviour when using their current central heating system formed the selection process to identify participants who reported distinct behaviour patterns. Deliberate efforts were made to identify participants from differing demographics and types of dwelling. Adjustments were made to the script following feedback from each participant such that each subsequent participant experienced an improved version.

The interviewer provided feedback as a 'participant observer' to evaluate the ease with which the QuACk prototype could be applied to the different case studies. The findings from both the case studies and observations were applied iteratively such that an improved version of the prototype was used on each successive case study. The resulting components and process which form QuACk are shown in Figure 3.4. Compared to prototype version depicted in Figure 3.3, this diagram contains 10 additional elements comprising verbal and textual positioning, question reviews, a section on terminology, instructions for the interviewer and a participant

TABLE 3.4

Bias Rings Identified in the Collection and Analysis of Data Derived from Interviews, Their Cause and the Mitigation Strategy Employed in the Development of QuACk Prototype

Identified Bias Ring	Cause of Bias	Strategy for Mitigation or Clarification
Background (participant and analyst)	• Previous experience with heating devices, formal knowledge of energy, thermodynamics, heating systems, attitude to using heating.	• Add questions and probes to gather data on background experience so responses can be viewed in this context. • Analyst answers questionnaire to clarify own background bias.

Social (means of communication)	• Participants' anxiety or embarrassment about incorrect, inappropriate or inconsistent answer.
	• Analysts' assumptions about meaning of responses, or accidently leading the participant.
		• To reduce anxiety and encourage free dialogue:
		  • Careful positioning of the purpose of the interview and the desired responses.
		  • Provide opportunity for participant to make changes to response.
		  • Provide opportunity for participant to assign confidence level to responses.
		• Follow recommendations by Oppenheim (1992) to reduce bias caused by interviewer when conducting structured interviews.

Cognitive artefact	• Pre-prepared terms or template for mental model construction biasing the responses given by the participant.	• No pre-prepared templates (e.g. blank paper, post-it notes, pen). • Terminology for components initiated by participant.

Social (method of analysis)	• Inaccurate data from which to form the basis of analysis due to leading, spontaneous construction of model during interview or misinterpretation of responses.
	• Data misinterpreted if analysed an extended time after the interview, or data are ambiguous.
	• Data unable to be compared and categorized if they contain different data elements.
		• To promote accurate interpretation:
		  • Record interview to help interpretation after extended time.
		  • Clear focus of content sought for outputs.
		  • Key outputs presented in a concrete, meaningful, unambiguous form to aid categorization without further transformation.
		  • Verification step by user when outputs complete to identify credible/less credible data.
		• To promote accurate data:
		  • Follow recommendations by Oppenheim (1992) to minimize leading.
		  • Paraphrasing to check meaning of responses to promote accurate data.

(Continued)

TABLE 3.4 (*Continued*)
Bias Rings Identified in the Collection and Analysis of Data Derived from Interviews, Their Cause and the Mitigation Strategy Employed in the Development of QuACk Prototype

Identified Bias Ring	Cause of Bias	Strategy for Mitigation or Clarification
		• Multiple opportunities for participant to make changes to outputs to reflect their true meaning.
		• Record interviews to allow posthumous analysis of spontaneous construction following techniques by Payne (1991).
		• Add probes during paper-based activities to encourage data elements required for comparison.

information sheet. The authors emphasize that these elements are treated by the interviewer as 'core elements' of the interview process rather than a 'nice to have' to ensure quality data capture. Appendix A provides the revised QuACk data collection method.

3.4 PARTICIPANT OBSERVATION – DATA ANALYSIS

With outputs in a form ready for analysis, the next requirement is a quick and clear process of analysis, which can be applied 'on-the-fly'. This section will (1) illustrate how the analysis reference table (Table 3.5) could be applied, (2) discusses the benefits of the form of outputs, (3) summarizes the utility of the analysis table as a means for categorizing home-heating mental models and behaviour patterns and (4) identifies changes to the analysis table that could improve its utility.

3.4.1 Applying the Analysis Reference Table

Participant 3 was highlighted to illustrate the analysis method, as they held a variety of model types and reported a variety of behaviour patterns (encompassing those depicted by participants 2 and 4). This provided an opportunity to illustrate a range of categories with one example.

3.4.1.1 Behaviour Pattern

Participant 3 was a 64-year-old female living in a four-storey Victorian terrace with her spouse. Both occupants worked full-time. The populated self-report template for typical behaviour when using the home-heating system is shown in Figure 3.5 (redrawn for clarity). The key in Figure 3.5 identifies three different devices used to control the heating (programmer, thermostat and boiler override). It also shows that these were operated by two agents (the participant and her spouse). The participant

TABLE 3.5

Analysis Reference Table for Quick and Systematic Analysis of Outputs from QuACk

Output	Criteria	Form of Output for Categorization			
		Valve	Feedback	Switch	Timer
1. Self-report of typical home heating use	Pattern of manual adjustment	Frequent and irregular when home is occupied *Influenced by events/ comfort*	Routine adjustments to thermostat *Event specific*	Set point is stable between events *Influenced by events/ comfort*	Higher settings remain unadjusted for longer periods than lower settings *Influenced by events/ comfort*
	Thermostat Set points	Varies above and below internal temperature	Specific temperatures	*Extreme high/low settings or 'until click' (i.e. 0.5°C above internal temperature)*	*Manual adjustments typically above internal temperature*
	Night set back	Yes	No	n/a	n/a
2. Mental model description of home-heating function	Thermostat set point and relationship to boiler	Variations in thermostat set point results in variations in intensity of boiler	Variations in thermostat set point inform the target room temperature to be maintained by the boiler	Increasing and decreasing the thermostat set point, activates or deactivates the boiler, respectively	Variations in thermostat control results in variations in the time period of boiler operation
	Boiler function	Boiler heats water to a range of temperatures	Boiler heats water to a single temperature	Boiler heats water to a single temperature	Boiler heats water to a single temperature
	Role of thermometer	n/a	Thermometer feeds back temperature value so comparison with thermostat set point can be made to determine if the boiler needs to come on or off	n/a	n/a

Note: Each output has key criteria corresponding to the form of output expected for each category of mental model held for thermostat function.

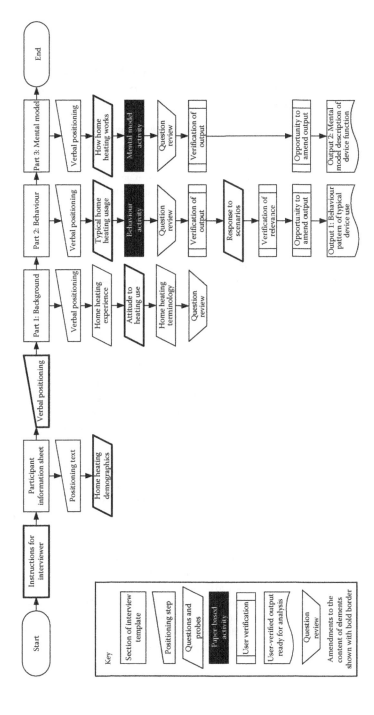

FIGURE 3.4   The key elements of QuACk following iterative development.

TABLE 3.6

Iterations to QuACk Resulting from Case Study and Participant Observations (Round Bullets Reflect Amendments to Method, Dash Bullets Identify Aspects that Worked Well)

Participant	Key Feedback and Amendments

Participant 1
51-year-old working male, living with a young family in a 4-bedroom detached bungalow with five different heating systems
	• Reiterate the purpose of the interview and expectations at the beginning of each section.
	• Add a 'terminology section' in the background section to define the elements in the function section as well as improve understanding.
	• Limit 'drilling down' to areas that are key for distinguishing between categories, or if the participant appears to be uncomfortable.
	• Verbalize during the attitude section that the study is non-judgemental and not aligned to a particular attitude about energy use, and improve the wording.
	• Change the wording of the analogy questions in the device function and system activities so a response is not mandatory, put at the end of each section.
	• For cases where multiple heating systems exist, focus interview on the main central heating system.
	• To avoid embarrassment from answering personal questions – (1) move demographic section out of the interview and provide in paper-based format for the participant to fill in prior to the interview, and (2) allow selection of age range instead of asking for an actual value.
	  • *Paraphrasing worked well and should be used throughout.*
	  • *The requirements of the self-report activities were easy to grasp after a few parts had been drawn out.*

Participant 2 25-year-old working female, living alone in the week, and with spouse at weekend *Session time: 30 minutes*	• Refer to target temperatures as 'a comfortable temperature' rather than providing a specific value, as people vary in their idea of a comfortable temperature. • At the start of the interview, verbalize to the participant that the interviewers' expertise was on data collection, not the workings of home-heating systems, so any verbal or facial cues would not relate to the accuracy of their answer. • Take care not to lead the participant when linking elements during the function activity. • Add a template to the 'self-report' of behaviour section, to speed up data collection.

(Continued)

TABLE 3.6 *(Continued)*

Iterations to QuACk Resulting from Case Study and Participant Observations (Round Bullets Reflect Amendments to Method, Dash Bullets Identify Aspects that Worked Well)

Participant	Key Feedback and Amendments
	• Add instructions to interviewer to review questions at end of each section. 　• *Reiteration of purpose of interview was welcomed by participant and not considered repetitive or patronizing.* 　• *Moving demographics to pre-interview questionnaire filled in by participant worked well.* 　• *Addition of terminology section at start of the interview enabled a more 'dynamic' tailoring of the interview script which improved understanding and engagement.* 　• *Change to analogy question removed pressure on participant to 'come up with something'.*
Participant 3 64-year-old working female living in a four-storey Victorian terrace with spouse *Session time: 75 minutes*	• Verbalize to the participant at the start and throughout that inconsistencies are normal and may be useful for research purposes. Take care not to demand consistency in participant responses, by referring back to things they have said previously, that contradict new responses. • Produce instructions for the interviewer at the start, and throughout the interview, to help keep on track. 　• *Instructions to interviewer at points within script were helpful for keeping on track.* 　• *Self-report template was a good format for recording behaviour with a variety of devices.* 　• *Self-report template was a good format for recording how many people in the home control the heating, which devices they tend to use and when they tend to use them.*
Participant 4 87-year-old retired male living alone in a semi-detached house. *Session time: 92 minutes*	• Advice in instructions to interviewer, that older participants have specific attributes that may require an adjustment in interview style, e.g. may wander off subject and require steering back to the question, are likely to talk about temperature in °F rather than °C, may respond to questions about home heating function as if they are expected to 'teach' the interviewer rather than 'explain' their understanding. • Advice in instructions to interviewer, that different degrees of knowledge about will affect the interview time. • No further changes to the format or approach of QuACk. 　• *Instructions to interviewer were useful for orienting and keeping on track.*

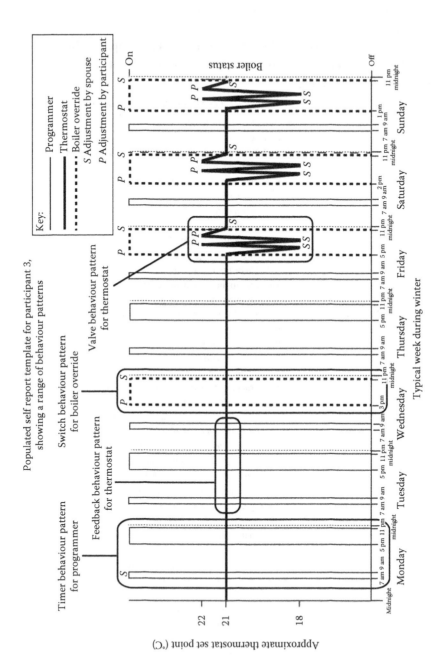

FIGURE 3.5 Output 1 from QuACk, redrawn for clarity, showing 'behaviour when using home heating over a typical week'.

was not sure of exact set points for the thermostat so the scale displays 'approximate' temperature values.

Referencing Table 3.5, and interpreting the contents for multiple control devices, different behaviour pattern categories have been highlighted in Figure 3.5. The most distinctive pattern is a 'valve' category assigned to the thermostat during the weekend. This pattern complies with the criteria in Table 3.5 as there are frequent, irregular manual adjustments when the home is occupied, with the set points varying above and below the (presumed) internal temperature. It is not clear, however, if night set-back is adopted as adjustments at the end of the day were undertaken by the spouse, so the participant could not recall the set point chosen. During the working week, the thermostat set point remains at a single setting, and the programmer is responsible for routine periods of boiler activation based around events such as returning from work. Table 3.5 does not consider how the behaviour pattern of the thermostat is affected by the use of a programmer. However, the lack of manual adjustment suggested a deliberate single temperature that implies 'feedback' behaviour, if it is assumed that the 'routine manual adjustment' described in Table 3.5 is taken care of by the program schedule. When completing the template, participant 3 described regularly using the manual override on the boiler, particularly when returning from work early and finding the house cold, or at home over the weekend. The participant switched the boiler setting to 'on', and this remained for a number of hours, until the spouse turned it off at the end of the day (see Figure 3.5). Interpreting the contents of Table 3.5 for the boiler override control, we can categorize this as a 'switch' model, as the behaviour pattern depicts a set point that is stable between events, and extreme set points (the only set point options for this control device). Figure 3.5 shows 'timer' category assigned to the behaviour pattern for the programmer. This cannot easily be inferred from Table 3.5's description of the 'pattern of manual adjustment' of the timer category for the thermostat. However, looking at the criteria for timer shared theory in Table 3.5 relating to the mental model description, the sense that variations in the settings resulted in variations of the timer period of boiler operation, was compatible to how the program settings were chosen.

3.4.1.2   Mental Model Description of Home-Heating Function

The mental model description of home-heating function for participant 3 has been redrawn for clarity (Figure 3.6), and depicts a range of control devices including the thermostat, programmer, boiler override and thermostatic radiator valves (TRV – described in Figure 3.6 as a 'radiator knob'). Each of these devices has been categorized using the prototype analysis reference table (Table 3.5). Participant 3 was distinctive in describing the function of the thermostat differently as the interview progressed, resulting in three descriptions and categorizations for the thermostat. This characteristic provides an opportunity to discuss different ways of categorizing the thermostat from one mental model description, as well as drawing attention to the reader that a participant may have multiple, contradictory models for a single device. A switch category from Table 3.5, was assigned as participant 3 described how the boiler comes on if you turn the thermostat until it 'clicks' (Figure 3.6). This matches the statement in Table 3.5 relating to the thermostat set points (in the self-report of

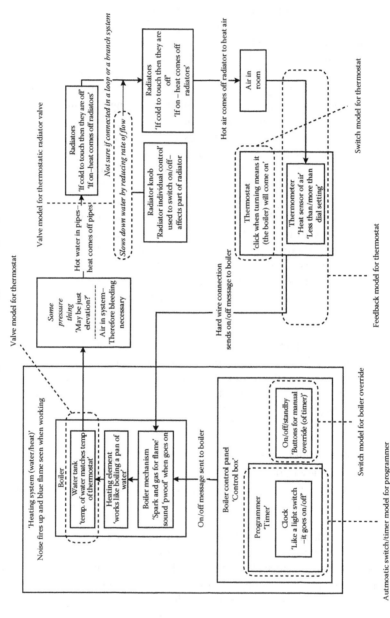

FIGURE 3.6 Mental model description of device function for participant 3, redrawn from output 2 for clarity.

behaviour section). It also conforms to the criteria for the relationship between the thermostat set point and the boiler for a 'switch' model. A feedback category was assigned to the thermostat due to: (1) the presence (in Figure 3.6) of a thermometer, (2) its described function to sense the air temperature in the room which is compared by the thermostat to see if 'less than/more than the dial setting', followed by (3) the link to the boiler with the rule 'to send on-off message to the boiler'. This describes the thermostat set point and relationship to the boiler described in Table 3.5. Table 3.5 requires that for the boiler function in feedback or switch models water is heated to a single temperature. This wasn't met in Figure 3.6, however. When discussing the boiler – a valve model of the thermostat was indicated – participant 3 described 'the temperature of the water [in the boiler] matches the temperature of the thermostat' (Figure 3.6). This fits Table 3.5's requirements for a valve model in terms of the boiler function, as well as the relationship between the thermostat set point and the boiler (Table 3.5), although the presence of the thermometer does not comply with 'valve' shared theories.

The mental model of device function is more consistent for the programmer and boiler override – both devices that participant 3 depicted in their self-report of behaviour output. Figure 3.6 shows that the programmer is referred to as a 'timer' and has a 'clock' component to determine when the boiler comes on. From Table 3.5, we see the relationship between the thermostat and the boiler for a timer model, requires variations in the thermostat control (or in this case, programmer settings) to result in variations in the time period of the boiler operation. This led to the categorization of the user holding a 'timer' model for the programmer. However, participant 3 also uses the analogy 'like a light switch, it [the boiler] goes on and off'. Whilst this terminology may seem to indicate the 'switch category', that this description is part of the 'clock' element of the model points to the fact that the clock is driving the activation/ deactivation of the boiler, rather than the user. This reconfirms the 'timer' categorization. As with analysis of output 1, by making the substitution of the control device in the description provided in Table 3.5, it is possible to infer relevant shared theory types for different devices. In Figure 3.6, the boiler override is simply described as an 'on/off/standby' switch, and is linked to, and has the capacity to activate/deactivate, the boiler. Again, substituting the control device in the description in Table 3.5, from thermostat to boiler override, is equivalent to the 'switch' shared theory description for the relationship between thermostat set point and boiler. Table 3.5 also requires, however, that for users with a switch theory, the boiler should function by heating the water to a single temperature (which was not evident in Figure 3.6). However, the variations in boiler temperature are clearly shown in the diagram as relating to the thermostat, not other control devices, so a switch categorization is still valid for the boiler override.

Figure 3.6 also depicts participant 3's description for the function of thermostatic radiator valves. They describe the purpose of the radiator knob to 'switch [the radiator] on/off', suggesting a 'switch' category; however, the transcripts indicate that this description reflects how participant 3 uses them, rather than her belief about their 'intended' purpose. The lack of thermometer element depicted by the TRV, and the rule describing how this (shown on the link between this control and the radiator in Figure 3.6) 'slowed down the water [to the radiators] by reducing the rate of flow' can be inferred from Table 3.5 to suggest a 'valve category'. By substituting the control device, the affected component and the variable, the description intended for the thermostat control's relationship to the boiler could be rephrased as 'Variations in the *radiator knob* [control device] set point results in variations in the *water flow* [variable] of the *radiators* [affected component]'.

3.4.2 BENEFITS OF OUTPUT FORMATS

3.4.2.1 Self-Report Diagram

The format of the self-report template had a number of benefits. Firstly, it was flexible enough to incorporate in a single view a variety of behaviour patterns from a range of control devices. This would be difficult to achieve from existing automated data collection solutions. Secondly, the ability to assign different agents to different aspects of the behaviour patterns was illuminating. From the perspective of exploring mental models, the 'valve' pattern shown in Figure 3.5 clearly represents a 'conflict of setting choice' rather than considered as continuous adjustment to ensure a comfortable house temperature (as described by Kempton 1986).

3.4.2.2 Mental Model Description

The mental model description is useful at identifying misunderstandings about how the home-heating system functions in a way that is concrete and explicit. For example, key elements of the heating system may be missing, elements may be inaccurately linked and the rules of cause and effect between elements may be incomplete or inaccurate. Revell and Stanton (2014, 2015) provide examples of in-depth analyses of mental model descriptions created by the QuACk method. They also explain how the insights gained can inform strategies intended to encourage behaviour change, or reduce energy consumption through the use of home heating.

The case study of participant 3 described an inconsistent mental model of the thermostat, yet clear and consistent (though not necessarily accurate) representations of other control devices. The ability of this output to identify inconsistent mental models has benefits. Inconsistent descriptions of function may be symptomatic of ambiguity in the communicated function of devices, or the relationships between devices. Where the ambiguity has negative consequences, in terms of performance, usability, or, in this case, wasted energy consumption, identification of this ambiguity could point to design or instruction strategies to clarify function.

3.4.2.3 Association between Mental Model of Device Function and Behaviour

The results of the analysis of verified outputs for participant 3 show a range of behaviour patterns and a range of mental models held. Purely based on the categorization, there is evidence of both valve and feedback mental models and behaviour patterns for the thermostat. This may suggest an association between behaviour and mental models for this participant. The case study presented described behaviour patterns that, unlike Kempton (1986), included other control devices (e.g. boiler override, programmer, thermostatic radiator valves [TRVs]). Any exploration between behaviour patterns and mental models of devices requires the same device to feature in both outputs, which was not always the case in the pilot study. However, as output 1 only considered 'typical' behaviour over a week, irregular behaviours, such as adjustment of TRV's, may occur less than this, explaining their absence. To consider an association with irregular behaviours, the format of the template to record behaviour would need to be adjusted.

Caution should be applied to conclusions regarding association from these outputs. For example, as evident from multiple agents being identified on output 1 from the case study, the valve 'pattern' does not necessarily reflect valve behaviour. If viewed through automated data collection, without assignment of agents for different aspects of the behaviour patterns, it could have been mistaken as such. Another example was seen in the categorization of the TRV in output 2. The transcripts indicated that the participant used the device as a 'switch', even though they thought the device functioned like a 'valve'. User behaviour with this control device was not primarily caused by their mental models of device function. The source of behaviour may represent a workaround to achieve goals that could not be achieved with other control devices, or to achieve goals that were not an intended function of the home-heating system (such as quick response individual room heat control).

The ability of output 1 to indicate multiple agents also helps inform further exploration of association. In Figure 3.5, the spouse was marked as responsible for the behaviour pattern relating to programmer. So, to identify an association, the mental model description of the programmer function would need to be provided by the spouse.

Evidence in both outputs indicated switch category for the boiler override and timer category for the programmer for both behaviour and mental models. This provides support for the relevance of existing 'shared theories' in the literature for categorizing models and behaviour associated with home heating. However, the authors wish to make clear that this is considering the shared theories from the literature in a 'generic' sense, rather than thermostat-specific.

3.4.3   Evaluating the Utility of the Analysis Reference Table

By applying the analysis reference table (Table 3.5) on the outputs of participants 2, 3 and 4, some key insights were found that recommended improvements/adjustments. These have been tabulated in Table 3.7.

3.4.4   Improvements to the Analysis Reference Table

Following the recommendations from Table 3.7, separate tables were created for categorizing behaviour and mental models. A generic example was provided for each shared theory, and specific examples from the literature relating to the thermostat were re-phrased to fit within the generic structure (see Appendix B). The intention was to make it easier to infer shared theories from alternate devices to the thermostat.

TABLE 3.7

Summary of Evaluation of Analysis Reference Table

Insight from Trailing Analysis	Suggested Change to Analysis Reference Table
All sections of the table were helpful when categorizing each output, suggesting the type of information captured extended beyond those expected for each output when the table was constructed. This may mean mental models data are being used to categorize behaviour data (or vice versa), or that the table doesn't distinguish effectively between the two types of data.	Make clear the key data to be analysed on each output, so that categorization can be attributed to *either* behaviour *or* mental models. Divide table into 2, so the relevant table is referenced when analysing each output.
Descriptions developed to categorize models relating to the thermostat control only. A range of control devices were evident on the outputs, requiring translation of the descriptions to a different device.	Make analysis table more applicable for outputs resulting from present-day heating systems in the UK. e.g. write descriptions to categorize model types generically, with thermostat as example.
Descriptions developed to categorize behaviour patterns relating to the thermostat only. A range of control devices were evident on the outputs, with different degrees of freedom, range and type (e.g. discrete, continuous, categorical) of input available. The expected behaviour pattern had to be inferred by the analyst, and might not be clear for all devices.	Make analysis table more applicable for outputs resulting from present-day heating systems in the UK. e.g. write descriptions to categorize behaviour types generically, with thermostat as example.
Inference of descriptions to other devices was aided by substituting either the control device, controlled element, and the change variable.	Define what is meant by these terms and use them (and others, if needed) to form the 'generic' descriptions for the table.
Outputs categorized with the same name of shared theory, but applied to different control devices represent a variation on those in the literature. The implications for behaviour patterns and the importance in terms of performance or energy consumption may differ.	Reinterpret the shared theories of home-heating use from the literature, within the framework of the generic categorization descriptions.
Sometimes data in the outputs applied to some of the category descriptions but not all. Not clear from the table if particular characteristics had greater weighting over others.	Make the table more 'granular' in its breakdown of characteristics required and provide guidance on 'critical' and 'supporting' characteristics.
Evidence of a mental model description with the model for a device described in variety of ways. This led to multiple, conflicting categorizations.	Provide guidance that multiple categorizations of a single control device are a useful finding. It identifies where the user has an inconsistent or incomplete mental model.

The process for categorization of outputs involves the analysts considering each control device (e.g. thermostat, programmer and TRV) separately and performing the following steps:

1. Identify and describe each of the elements from the table heading (control device, input behaviour, key element, key variable, sensor/sensed variable, rule).
2. Compare the descriptions to the generic/shared theory descriptions in the table.
3. Select most appropriate category or identify variations to closest category.

To aid this process, question sheets corresponding to each output were created, with examples of typical answers within the home-heating context (provided in Appendix B). This version of the analysis reference tables, as well as the question sheets were used in the reliability exercise described in Section 3.5.2.

3.5 VALIDATION

Method validation is a key attribute of a good method; however, Stanton et al.'s (2014) compendium of human factors methods reveals that this step is often over-looked. As the aim of QuACk is to allow exploration of association between user behaviour with home-heating systems and their mental models of those systems, an understanding of how well output 1 captures actual behaviour is clearly important, as well as the ability of the method of analysis to consistently categorize outputs. This section describes initial attempts to assess measurement validity of output 1, and the reliability of output categorization.

3.5.1 Measurement Validity of Self-Report Behaviour

The validation approach was focused around the criteria used in the analysis table since the emphasis was to seek validation of behaviour characteristics relevant to users' 'shared theories' or 'generic models' of home heating.

Due to the nature of the home-heating domain, direct observation of home-heating behaviour over week-long periods of time was impractical (as discussed in Section 3.2.2). To seek initial validation of output 1, the spouse of the user (where applicable) was contacted to see if they agreed with the representation provided in output 1 for their dwelling by the user. The spouse was asked to check the set point values of control devices represented on the output (e.g. programmer schedule times and thermostat set points). The author then showed and explained the pattern on output 1 and asked the spouse to express their level of agreement that this reflected their household. The authors annotated the output to capture their responses, and tabulated the results (Table 3.8).

Table 3.8 shows that, overall, spouses were in broad agreement that output 1 represented their typical weekly behaviour with the home-heating system. There was full agreement for the control devices used, and general agreement with the distribution between agents, frequency, regularity and synchronicity of the pattern, as well

TABLE 3.8

Summary of Spouses' Agreement with Behaviour Shown in Output 1 (Agreement = ✓, Disagreement = ✗)

Agreement by Spouse[a] for Features of Output 1	Participant A	Participant B	Participant C[a]
Devices used	✓ only boiler override used	✓ programmer, boiler override and thermostat typically used	✓ programmer and thermostat typically used
Set points chosen Variations in set points	✓ boiler override set to either on/off ✓ durations and times for boiler 'on' periods	✓ number and duration of sessions for programmer accurate ✓ timings for programmer reasonably accurate (±30 minutes) ✓ set points for thermostat broadly correct (±1°C) ✓ variations in thermostat set point	✓ programmer set points times and durations accurate ✓ static thermostat set point during week, and significant reduction when out for the day ✗ thermostat set point value inaccurate (actual = 21°C, output 1 shows 90°F = to 32°C)
Regularity, frequency and synchronicity of pattern	✓ weekend pattern typical (could not verify weekdays, as absent from dwelling)	✓ weekday pattern ✗ extended time for override at weekend occurs typically on one, rather than both days	✓ weekday and weekend pattern
Distribution of agents across pattern	✓ user responsible for deactivating boiler override at weekend ✗ spouse OR user activates boiler at weekend (not spouse alone)	✓ spouse responsible for setting programmer ✓ user raised thermostat set point and activated boiler override and spouse reduced thermostat set point and deactivated boiler override	✓ sole user of devices, responsible for all adjustments

[a] Participant C was a lone occupant. He was asked to check device settings and give views on output 1 that he had constructed himself, 1 year after the original interview.

as set point values and variations. There were some exceptions, however. Participant C revealed a large inaccuracy with the thermostat set point value, though, this is likely to be a conversion error between units of temperature. There was also minor disagreement in terms of regularity of pattern (participant B), and distribution of agents (participant A), though these were not sufficient to alter a categorization of the pattern.

The results of this initial validation was positive in terms of the value of output 1 as an appropriate measure to categorize behaviour according to that predicted by shared theories or generic theories. If part 1 of QuACk was used to generate output 1 for other research goals, such as estimating energy consumption, or understanding the assignment of agents to behaviour, further checks may be needed if a high level of accuracy is required

3.5.2 Reliability of Analysis Method

To test the inter-analyst reliability of the updated analysis method, two human factor analysts were asked to analyse three sets of output resulting from interviews performed using the QuACk method. Their results were compared with the authors' own categorization of this data to evaluate their level of agreement. The analysts were provided with: (1)'walk-through question sheets' (with examples) that guided observations of the data, (2) 'answer sheets' for recording responses from the question sheet in a form that could easily be compared to the reference, (3) the analysis reference tables for each output, (4) outputs 1 and 2 from each participant and (5) transcribed 'paraphrases' describing each output, taken from the interview transcripts. The paraphrase was provided to help orientate the analysts to the outputs. This was in recognition that an analyst using QuACk would typically have interviewed the participant and constructed the outputs, so would be fully orientated to their meaning at the point of analysis.

3.5.2.1 Dynamics of Exercise

Prior to analysis, the author trained the analysts in the analysis method. The outputs of participant 3, categorized with generic categories from the updated analysis tables, were used as an example and the analysts were walked through how to use the walk-through question sheets, answer sheets and analysis reference tables. It was typical of output 2 (mental model descriptions) to have more control devices evident than output 1 (behaviour patterns). As the goal of QuACk is to explore association between behaviour and mental models, the analysts were asked only to categorize control devices on output 2 that were present, and operated by the participant, according to output 1. This resulted in six devices needing categorization for each output type. The validation exercise ran for 2 hours, with the training session taking up approximately 30 minutes and the categorization approximately 1.5 hours. The mean time for categorization of a single device from a single output was around 10 minutes. The analysts were only allowed to ask questions to clarify analysis method, during the exercise, but were not given guidance on the appropriate category to select.

3.5.2.2 Results of Inter-Analyst Reliability Exercise

Table 3.9 summarizes the results of the inter-analyst reliability exercise. This table shows the level of agreement with the authors' analysis of the same outputs using the updated analysis tables and walk-through questions.

Table 3.9 shows that overall agreement with the categorization stood at 79%, which was reasonable for analysts who had limited orientation to the outputs – 80% is considered an acceptable level of reliability (Jentsch and Bowers 2005). Agreement levels in the categorization of output 2 were very good, with 92% agreement, validating the utility of analysis reference table for categorizing mental model

TABLE 3.9

Results of Inter-Analyst Reliability Exercise

Participant	Devices	Output	Author Categorization	Analyst 1 Agreement	Analyst 2 Agreement
Participant A	Programmer	1 – Behaviour pattern	Generic FB1	✗ Timer Norman (2002)	✓
		2 – Mental model desc.	Generic FB1	✓	✓
	Override	1 – Behaviour pattern	Generic switch	✓	✓
		2 – Mental model desc.	Generic FB2	✓	✗ Generic valve
Participant B	Thermostat	1 – Behaviour pattern	Generic FB1	✗ Valve Kempton (1986)	✓
		2 – Mental model desc.	Generic FB1	✓	✓
Participant C	Programmer	1 – Behaviour pattern	Generic FB1	✓	✓
		2 – Mental model desc.	Generic FB1	✓	✓
	Override	1 – Behaviour pattern	Generic switch	✓	✓
		2 – Mental model desc.	Generic switch	✓	✗ Generic valve
	Thermostat	1 – Behaviour pattern	Generic FB1	✓	✗ Generic valve
		2 – Mental model desc.	Generic valve	✗ Generic FB1	✓
% Agreement	Categorization	1 – Behaviour pattern	67%		
		2 – Mental model desc.	92%		
		Overall	79%		

descriptions. Analysis reference table for output 1, however, elicited lower levels of agreement with the authors' categorization, at 67%. After examining the completed answer sheets provided by the analysts for the thermostat patterns for participants B and C, the authors found the following causes of disagreement. Analysts tended to categorize 'valve' for thermostat if reference was made in the transcribed paraphrase that set points were based on 'comfort'. The analysis table explicitly describes the generic feedback category for the thermostat as based on lifestyle activities rather than comfort (taken from Kempton 1986). However, this does not imply that comfort is not the desired goal from the set point choice, only that constant adjustments to the thermostat are not made to compensate for changing comfort levels from other causes (e.g. activity levels). This distinction was not sufficiently clear in the table. The label of the category may also have encouraged a different categorization. Analyst 1 assigned a 'timer' category (intended only for thermostat devices) to the programmer. As the programmer is often known as a 'timer', and the description refers to variations in time, this association may have seemed more similar than the generic feedback 1 category. In addition, the descriptions in the two feedback categories may not have emphasized their distinguishing features. Only 1 mis-categorization of output 2 depicting the mental model description was found with the thermostat device for participant C. The analyst's response to 'Q1c – description of automatic adjustments' showed they misunderstood the meaning of the question. They provided the example 'it sends messages to the boiler' as evidence of an automatic adjustment. However, this depicts a step in a process, not an adjustment. Further training or a clearer distinction in the question could avoid this in the future.

3.5.2.3   Improvements

Providing more examples and further explanations of the walk-through questions and examples are likely to improve categorization. Using analysts who have conducted interviews and produced the outputs for the validation exercise may also result in faster categorizations. In terms of the analysis tables, a greater emphasis of the distinction between categories, either textually or with diagrams, would also help.

3.6   DISCUSSION

This chapter demonstrated how the development of QuACk considered key methodological issues in ergonomics science; how, through a qualitative case study approach to iterative development, a more robust method resulted, and discussed the potential of applying the findings of QuACk to energy-reducing strategies.

Kanis (2002) stated that the way validity and reliability is assessed in human factors research is often lacking or misconceived. This chapter has illustrated the ways in which the development of QuACk has taken steps to pursue what Stanton (2002) considers a hard-won, but worthy goal. QuACk has three stages in application: (1) behaviour pattern capture (2) mental model capture and (3) categorization of outputs. Initial reliability tests were undertaken by asking the spouses of pilot participants to verify how well the behaviour patterns described reflect their own perception of the participants' behaviour. Broad agreement was found regarding the behaviour patterns adopted, though some variations were found in the value of specific thermostat and programmer set points. For pattern categorization based on part 3 of QuACk, this is encouraging, but alternate analysis of QuACk outputs that relied on set point value accuracy would benefit from automated data collection methods (e.g. as used in Revell and Stanton 2015). The greatest challenge presented to the authors was pursuing the reliability and validity of mental model outputs. The reliability of mental model outputs, in terms of a technical sciences definition, whereby the same output would result from reapplication of the method (Kanis 2004), is something of a 'misnomer' in mental models research, as mental models are dynamic constructs that can change in content over time (Johnson-Laird 1986). Whilst triangulation is often used to validate data, Kanis (2004) argues that this treatment is often flawed as assumptions need to be made about the link between the outputs of different methodologies. This criticism could be levied against mental models research that emphasize the face validity of a causal link between user mental models of device function and behaviour with devices (Gentner and Stevens 1983). However, an analysts' description of a user mental model is not equivalent to the knowledge construct itself (Wilson and Rutherford 1989). The authors adopted the position of Annet (2002), who distinguishes between analytic and evaluative methods. QuACk falls into the criteria of an analytic method, as it produces a better understanding of processes affecting complex human–machine systems, such as the domestic heating system (Sauer 2009). Annet argues the validity of analytic methods is based on its relationship to underlying theory. In this chapter, the authors have demonstrated a number of steps that ground QuACk in mental models theory. A key differentiator from other methods is that the development of QuACk considered bias at its outset an important risk to validity in mental models research (Rouse and Morris 1986, Wilson and Rutherford 1989, Bainbridge 1992, Richardson and Ball 2009, Revell and Stanton 2012). Section 3.2.4 emphasizes the many strategies that were put in place to mitigate for unwanted bias through the tree-ring method (Revell and Stanton 2012), such as verification of outputs by the user, format of outputs ready for analysis, paraphrasing and opportunities for amendment throughout the interview, provision of interview script and instructions to minimizing leading, and so on. For the domain for investigation, theory grounded constraints were included in the choice of research and probes (Section 3.2.2.1), the categories of shared theories for the analysis table (Sections 3.2.3 and 3.2.6) and the format of outputs (Section 3.2.5). These constraints were necessary to target the specific research question, and allow meaningful comparison between participants, to fulfil the need of practitioner to quickly explore association in a defined domain (Desnoyers 2004). To assess QuACk's reliability in interpretation of outputs, an inter-analyst-reliability test was undertaken. This showed good overall level of agreement with the authors' own categorization, very good agreement with categorization of output 2 (mental model description) and moderate agreement with output 1 (behaviour pattern).

The second aim was realized in Section 3.4 in two stages, considering, first, the prototype interview script and data collection method and, second, the utility of the analysis reference table. The results and recommendations from pilot studies and participant observation that related to the interview script and data collection method were tabulated, and the change in components was depicted visually in Figures 3.2 and 3.4 (where the components of the interview script before and after amendments were shown). To demonstrate the benefit of participant observation on the development of the analysis reference table, the outputs from one of the pilot participants was analysed using the prototype reference table (see Section 3.4.3). The challenges and insights gained from this process were tabulated and led to the production of two separate 'generic' analysis tables, corresponding to either the behaviour or the mental model outputs.

The development of QuACk using an iterative case study approach revealed issues fundamental to exploring association between models and behaviour of home heating. These include the consideration of multiple control agents, and the use (or failure to use) of multiple control devices. The qualitative approach for method development inspired by Hancock and Szalma (2004) was invaluable in the iterative development of QuACk. The authors found that the improvements to the interview script and template, resulting from the qualitative iterations, largely related to positioning, guidance and structure. It is these sorts of improvements, the authors believe, that can only be gained from qualitative methods, as 'hard and fast rules' for specialized contexts may be lacking. Through the participant observation evaluation of the analysis reference table, consideration of the way the authors 'inferred' the characteristics of shared models from the literature to alternate devices was insightful. This led to a 'generic' way of looking at shared models of home heating, which accommodated shared theories from the literature as specific examples. This supports Hancock et al.'s (2009) sentiments that generalities can be extracted from single case studies. The initial reliability tests showed good agreement between self-reported behaviour and actual behaviour, in terms of the impact on categorization of the data.

3.6.1 METHOD EVALUATION

QuACk was developed to fulfil the need of a quick, resource-light method for exploring association between mental models of, and user behaviour with, domestic heating systems. Further, the method was to consider bias in its development, and allow identification of shared theories of home heating that exist in the literature.

Section 3.4.2 describes the benefits of the behaviour and mental model outputs from QuACk, which are offered either only in part, or are absent in existing methods used for mental model and behaviour capture, such as verbal protocol analysis, critical decision method probes, cognitive walk-throughs, automated data collection of set points and user behaviour diaries. For example, QuAcks outputs provide concrete representations of home-heating models and behaviour, which can be analysed by practitioners without further transformation. They provide a flexible format to record key information relating to behaviour that present challenges to automated data collection (such as the number of agents responsible for behaviour and range of control devices used). They highlight behaviour patterns that arose from combinations of devices (e.g. if the thermostat is used in combination with a programmer, 'expected' behaviour patterns for users who hold a feedback mental model would differ to the 'expected' patterns if the thermostat was used alone) and, finally, they are able to reveal not only misunderstandings or misuse of heating systems, but inconsistencies and ambiguities in the participants' home-heating models and behaviour. With these characteristics, the outputs from QuACk offer the opportunity, not easily found with other methods, not only to focus further research, but to identify novel models and behaviour characteristics that could enhance the understanding of home-heating use in the United Kingdom. Collins and Gentner (1987) and Norman (1983) recognized variations in the completeness of mental models between individuals which could account for findings relating to misunderstandings, misuse and inconsistencies in models held (such as with participant 3's multiple models of the thermostat device). This characteristic could be interpreted in line with Johnson-Laird (1987) to suggest that the device itself is ambiguous in the way it presents its function. Multiple or conflicting models could then be considered 'symptoms' of issues such as ambiguous design, insufficient instruction or misleading feedback at a 'system' level. Further work by Revell and Stanton (2014, 2015) has shown the utility of the QuACk method to explore these issues. Revell and Stanton (2016) have, in addition, applied QuACk to investigate Norman's (1986) gulf of evaluation and execution between users of home-heating devices and the intended operation by manufacturers. This work produced a 'design spec' based on mental model analysis to direct the design of home-heating systems in a way that better allows user to produce an energy-conserving 'action specification' to achieve their home-heating goals. Salmon et al. (2010) considered the landscape within which methods may contribute to system design and evaluation. The authors consider QuACk sits towards the design modification stage, as it highlights misunderstandings with the functioning of current systems that could direct a designer to make the functioning more explicit.

Whilst QuACk has been developed specifically for the home-heating context, the format could be adopted for exploring association between user mental models and their behaviour with other devices or systems. The mental model activity, like a concept map, is not domain-specific, and the self-report of behaviour template could be applied to devices where behaviour patterns are based on set point choices over time. The authors have already used the method to access user mental models of a novel heating interface displayed in a home-heating simulator, as well as adapted the wording to understand pilots' mental models of hydraulic failure in the cockpit and associated behaviour. Researchers or practitioners who were interested in closely related domains, such as non-domestic heating systems, air-conditioning/cooling systems or domestic hot water systems, would need only make minor variations to the wording and content of questions, and scale of behaviour output. The authors believe there may be further applicability to domains other than space heating. For example, domestic energy-consuming devices that present simple input controls, but do not have a clear relationship to the task being performed or energy consumed (e.g. microwaves, washing machines, tumble dryers, electronic devices). For these devices, adjustments to the content of the questions and the scenarios presented would be needed. The format of the template for recording self-reported user behaviour would need to be based on either existing behaviour patterns identified in the literature, or follow from exploratory data collection methods such as observations, participant observations or in-depth case studies. With application to a context outside of domestic home heating, the authors recommend iterative development of the questionnaire and analysis table (see Section 3.4). Stanton et al. (2005) consider the information necessary to communicate a human factors method effectively comprise its background and applications, domain of applications, procedure, related methods, application times, advantages and disadvantages, reliability and validity, as well as the tools needed. This chapter encompasses these criteria in the presentation of the QuACk method.

3.7 CONCLUSIONS

Wilson and Rutherford (1989) emphasize that applying the notion of mental models (particularly in design) increases the total effort to the practitioner. To ensure the benefits of using this notion, in terms of performance or comprehension, they demand that rigour in determining the mental model description is necessary. The authors offer that the systematic process in developing QuACk and analysing the output is a step in the right direction. Whilst initial outlay is required in the development of the method, the reduced time in data collection and analysis that results from a structured interview yielding verified outputs 'ready for analysis' and in a concrete, meaningful form for the practitioner, furthers this goal.

The QuACk (quick association check) method was developed using qualitative approaches and is grounded in theory to improve construct validity. It is a structured interview method which includes activities and templates to produce verified outputs ready for analysis. QuACk also provides analysis reference tables for each output, and 'walk-through' questions to guide the analyst. The benefits of the method include flexibility, speed of data collection and ease with which data can be analysed to explore association. The benefits of rich data from the interview transcripts provide insights which can explain phenomena, target future research or determine appropriate strategies for mitigating inappropriate behaviour when operating devices. The authors anticipate that the method could be adapted to other devices and domains such as electronic consumer goods, dashboard and cockpit devices in transportation or control room design in military or nuclear domains.

4 Case Studies of Mental Models in Home Heat Control

Searching for Feedback, Valve, Timer and Switch Theories

4.1 INTRODUCTION

This chapter looks at the relationship between householders' mental models of home heating at both the device and system levels. It considers how differences in the mental model held can explain householders' self-reported behaviour with heating controls. This contributes to hypothesis 1 described in Section 1.4.1, which investigates the influence of users' mental models of devices on their pattern of device use. Greater understanding of the link between mental models and behaviour is critical to furthering this book' key aim, to use the notion in design strategies in order to elicit changes in behaviour that increase the achievement of home-heating goals.

Mental models can be thought of as internal constructs that explain human behaviour (Wickens 1984, Kempton 1986). The notion has been associated with many domains over the last 20 years, including domestic (Kempton 1986), transport (Weyman et al. 2005) and military (Rafferty et al. 2010). Mental Models have formed the basis of strategies to improve interface design (Carroll and Olson 1987, Williges 1987, Norman 2002, Baxter 2007, Jenkins et al. 2010), to promote usability (Norman 2002, Mack and Sharples 2009, Branaghan et al. 2011, Jenkins et al. 2011, Larsson 2012) and to encourage sustainable behaviour (Kempton 1986, Sauer et al. 2009, Lockton et al. 2010), amongst others. In 1986, Kempton described two distinct 'forms' of mental models of thermostat function that were prevalent in the population of that time. He proposed that the form of model held could result in significant variations in the amount of energy consumed due to home heating, by promoting different patterns of manual thermostat adjustment. Currently in the United Kingdom, 25% of carbon emissions are from domestic customers, 58% of which is due to domestic heating. The United Kingdom has legislated to cut 80% of greenhouse gas emissions by 2050 (Climate Change Act 2008). Since Kempton's study, almost three decades have passed and technology has changed. It seems appropriate, therefore, to explore if Kempton's (1986) shared theories can still be identified, and if so, to question if they remain relevant to design strategies targeted at combatting climate change.

As described in Chapter 2, the term 'mental model' is used in different domains to mean different things (Wilson and Rutherford 1989) and, even within a domain, it can be used to describe internal constructs that differ significantly in terms of content, function or perspective (Richardson and Ball 2009, Revell and Stanton 2012). The form of mental model descriptions may have similarities to the way other types of models (e.g. process models or logic models) that do not depict internal constructs are represented, resulting in confusion when interpreting outputs. Specificity in the type of mental model is considered essential for commensurability when conducting research (Norman 1983, Wilson and Rutherford 1989, Bainbridge 1992, Revell and Stanton 2012). Please note the extended clarification of the way the term is used in this chapter. The intention is to allow sufficient understanding to determine the relevance and applicability of the findings presented. This chapter refers to mental models in three different ways: (1) in terms of its function, (2) in terms of its source and (3) in terms of its individuality.

In terms of function, the definition most fitting is a 'device model'. Kieras and Bovair (1984) adopted this terminology to describe a mental model held by a user of how a device works. It includes a set of conceptual entities and their interrelationships (Payne 1991). In this chapter, the device of interest is the home-heating system, and we seek to describe the conceptual entities and their interrelationships held by users. Device models, as a type of mental model, may be incomplete, inaccurate and inconsistent (Norman 1983). It is proposed, in this chapter, that understanding where omissions, inaccuracies and inconsistencies occur in users' device models of home heating could provide insights into how to reduce energy consumption resulting from non-optimal operation.

In terms of its source, this chapter adopts Norman's (1983) definition of a 'user mental model' (UMM). He describes this as 'the actual mental model [of a target system] a user might have', which can only be gauged by undertaking observations or experimentation with the user. In this chapter, we seek the model of the home-heating system held internally by a user. As we cannot access this model directly; therefore, we have adopted a method appropriate to our aims to gain data to describe the user mental model.

In terms of the individuality of mental models, we also refer to Kempton's (1986) 'shared theory'. A 'shared theory' is derived by an analyst through the identification of similarities in separate UMMs of individuals. These individuals are within a social group who may share similar types of individual goals. A 'shared theory' differs from concepts such as 'shared' or 'team' mental models that refer to shared knowledge structures within a team or group who are working towards group goals (Richardson and Ball 2009). The benefit of identifying shared theories of home heating is a broader reach when targeting strategies to combat climate change at individuals within the home.

The two shared theories identified by Kempton (1986) were described as 'valve' and 'feedback'. Users with a valve shared theory considered changes in the set point of their thermostat to be controlling the intensity of heat in their furnace, with the onus on the user to ensure a comfortable home temperature. Users with a feedback shared theory considered their responsibility merely to select the desired thermostat set point. The thermostat would maintain comfort in the home by controlling the boiler operation period, in response to measurements of house temperature. Kempton (1986) referred to this latter theory as an 'amateur theory' of home heating, as it is a simplistic version of the actual way the heating system works. Kempton (1986) described how different shared theories may predict different behaviour patterns of thermostat set point adjustment. He discovered that holders of valve theory had a unique behaviour characteristic, absent in those holding feedback theory. At night, valve theorists regularly set the thermostat back to below normal comfort levels, which Kempton (1986) described as 'night set back'. Kempton (1986) proposed that despite the valve theory being less accurate than the feedback theory, this behaviour characteristic was likely to result in greater energy savings overall.

Since Kempton (1986), additional shared theories of thermostat function have been proposed in the literature such as 'timer' (Norman 2002) and 'switch' (Peffer et al. 2011). Users holding the timer theory are thought to select greater values of set point when longer periods of boiler operation are desired. Those holding the switch theory are thought to use the thermostat merely as an on/off switch. Both of these theories assume the user, not the system, is responsible for maintaining a comfortable house temperature. Norman (2002) and Peffer et al. (2011) do not refer to studies which informed these types of shared theory, nor do they describe distinct behaviour characteristics which may influence energy consumption. When investigating current user mental models of home heating, it is relevant to determine if these, or new shared theories of home heating, could be identified. Understanding how resulting shared theories associate with energy-consuming behaviour could provide insights to inform novel approaches to reduce consumption.

The reader may question if mental models need to be accurate or is it sufficient that they are effective. Depending on context and the specific user behaviour being considered, what is considered 'effective' will vary. Kempton (1986) described how a faulty mental model of home heating control could lead to more energy-efficient behaviour than a more accurate model. Norman (1983) contends that designers and instructors should ensure a 'functional' (not necessarily accurate) mental model to enhance user interaction with a system. Norman (1986) emphasizes that the appropriateness of the user's underlying model of a system is essential when troubleshooting, as the user is able to derive possible courses of action and possible system responses. Kieras and Bovair (1984) concluded that for very simple devices or procedures, there will be little value in providing a device model to users. Manktelow and Jones (1987) warn that systematic errors may result from an inappropriately simple mental model. So, taken together, it is concluded that for simple procedures, simple devices or systematic errors that have minor consequences, a 'functional', simplified or even lack of mental model may be effective. For more complex systems or procedures, where the need for troubleshooting is likely, or if the consequence of systematic errors is significant (as in the case of non-optimal home heating during an energy crisis), a more accurate user mental model may be needed for the effective use of devices.

Hancock and Szalma (2004) emphasized the importance of qualitative methods in revealing user intention in a way that can inform the development of design principles. Flyvbjerg (2011) argues that rich data gathered from detailed, real-life situations can provide meaningful insights that could not be gained from context-independent findings. Virzi (1992), when conducting research into usability, found that 80% of problems, including the most severe, are detected with the first four or five subjects, illustrating how key insights can be gained with very small samples. Hancock et al. (2009) also argue that ideographic case representations are increasingly relevant for the design of human–machine systems, as advances in technology begin to focus on exploiting individual differences. Supporting these sentiments, this chapter describes the results from an inter-group case study of home-heating control, focusing in detail on three individual case studies taken from a pool of six. The aim of this chapter is to (1) demonstrate the existence of distinct mental model descriptions of the functioning of present-day UK home-heating systems, which differ significantly from actual functioning; (2) seek evidence of Kempton (1986), Norman (2002) and Peffer et al.'s (2011) shared theories of thermostat function in the case study group; and (3) discuss the present-day relevance of Kempton's (1986) valve and feedback models of thermostat function to design strategies targeted at combatting climate change. Additional implications and the limitations of the study are also discussed.

4.2 METHOD

4.2.1 Participants and Setting

The case study group was non-randomly selected and comprised mainly overseas postgraduate students with families, new to the United Kingdom, who resided in semi-detached university-owned accommodation in Southampton, UK. Participants arrived at their accommodation at the start of September 2011 and used the central heating system during the autumn and winter months. Southampton has an oceanic climate, with cool winters (temperatures typically below 5°C). The accommodation, home-heating devices and levels of insulation were matched, so that variations in mental model descriptions could be attributed to characteristics of the participant, rather than the environment. The layout of the home-heating devices and specific models used are shown in a diagram in Figure 4.1. The participants were recruited by letter, email and approached door-to-door by the author. Permission was sought from the faculty ethics committee prior to contact and research governance was arranged. The participants that agreed to take part were all from warm countries where centralized home-heating devices are uncommon. This user group characteristic, whilst not originally sought, ensured minor experience of other home-heating devices. This benefits the mental model descriptions of home-heating systems, by making them more closely aligned to the specific home-heating devices installed, rather than previous experience by the participants of other home-heating devices. Insights from this case study could therefore use the specific design and layout of the setup as a starting point for energy-saving strategies.

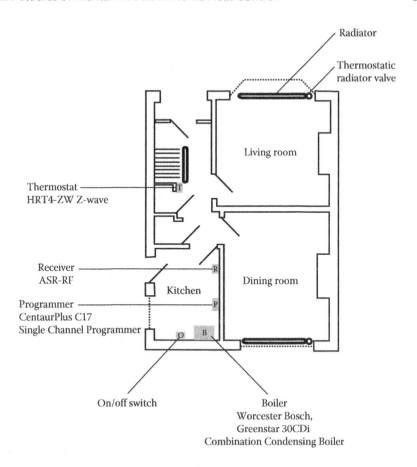

FIGURE 4.1 The layout of the home-heating devices and specific models used during the inter-group case study.

4.2.2 DATA COLLECTION

A pragmatic world view is adopted in this research. Whilst a post-positivist world view is suggested by the objective to verify Kempton's (1986) shared theories of home heating, to ensure the data collection method allowed interpretations of the data beyond this scope, it was important that alternate shared theories or unique UMMs could be revealed. This would allow further understanding of people's mental models of home-heating function to be gained, which could inform design-based strategies (amongst others) for reducing energy consumption. A method adapted from Kempton (1986) and Payne (1991) was developed to these ends. Payne (1991) described 'shared theory' device models in a concrete diagrammatic form that better communicates misunderstandings of function to design practitioners, than textual or non-deterministic schematic representations. Kempton (1986, 1987) used in-depth interviews and analysed the transcripts using metaphorical analysis (devised by Lakoff and Johnson 1981). He provides recommendations on the interview process and example questions and probes specific to home-heating systems. From this, an interview approach could be developed to access content that allowed the analyst to identify shared theories. The resulting method devised in this chapter was a semi-structured interview that included a paper-based activity whereby the participant represented their device model in a concrete diagram form.

For pragmatic application of mental models research, Revell and Stanton (2012) emphasized in Chapter 2 the importance of accuracy in the capture and representation of internal constructs. This accuracy requires the description of the mental model to reflect its source (in this study, the user of home-heating systems), rather than assumptions by the recipient (e.g. the analyst). The risk of bias as a means of causing inaccuracy in mental model research, and the need to take pains to minimize bias is well documented in the literature (Rouse and Morris 1986, Wilson and Rutherford 1989, Bainbridge 1992, Richardson and Ball 2009, Revell and Stanton 2012). In Chapter 2, Revell and Stanton (2012) developed a 'tree-ring' method in order to identify risk of bias when conducting mental models research, resulting in an adaptable framework (presented in table format). This framework required specification of both the risk of bias and the type of knowledge structure, to aid commensurability (see Revell and Stanton 2012 and Chapter 2, for an example of the tree-ring method applied to Kempton (1986) and Payne (1991) plus the resulting table). The tree-ring method (Revell and Stanton 2012 and Chapter 2) was applied to the approach devised for this study, and amendments were made to the data collection and analysis process in response to identified bias. The risk of bias identified through the tree-ring method (Revell and Stanton 2012 and Chapter 2) related to: (1) the background experience of both the analyst and participant, (2) the social expectations and means of communication of both the analyst and participant, (3) the structure of cognitive artefacts used in the interview and (4) the method of analysis of cognitive artefacts. These types of bias and the strategy for mitigation or clarification are shown in Table 4.1. The leftmost column shows criteria to be specified, with the section labelled 'perspective of gathering data' standard in every table. The section labelled 'layers of risk of bias' is specific to the approach considered. The type and number of layers of bias are dictated by the results of the tree-ring method. The middle column specifies the details required. Alphabetized in italics are types of relevant bias that may need mitigation. The rightmost column details how the approach adopted responded to the identified bias (also alphabetized). Table 4.1 allows readers to gauge the scope of efforts to promote accuracy in the capture of the UMM in the mental model description produced.

4.2.3 Dynamics of the Interview

The interviews were undertaken in March 2012, in a library cafe at the University of Southampton, to comply with the requirements of the risk assessment. The informal, familiar setting helped place subjects at ease. Interview durations were approximately 1 hour, but varied, depending on the level of detail provided by the participant, between 45 minutes and 1 hour 25 minutes. Interviews were recorded and transcribed by an independent transcription service. The interview comprised

TABLE 4.1

Risk of Bias and Mitigation Strategy for Method Adopted, Derived Using Tree-Ring Method from Revell and Stanton (2012) as Described in Chapter 2

Perspective of Gathering Data	Mental Models in Home-Heating Control	
Mental model description	Valve, feedback, as defined by Kempton (1986), switch (Pfeffer et al. 2011)	
Context (domain, behaviour/task, goal)	Domain: domestic	
	Behaviour: patterns of adjusting thermostat dial,	
	Goal: comfortable body temperature in a family home, reducing waste of energy/money	
Mental model definition used (e.g. shared model, user mental model, device model)	User mental model (Norman 1983)	
	Shared theory (Kempton 1986)	
	Device model (Kieras and Bovoir 1984)	
Source (e.g. device user(s), designer, analyst)	Users of domestic central heating systems in the UK	
Intermediaries (e.g. analyst(s), system image, instructor)	Analyst	
Recipient (e.g. analyst, designer, academic community)	Academic Community interested in climate change, behaviour change, device design, psychology, human factors	
Layers of 'risk of bias'	Inter-group case study	Mitigation strategy in method to address bias identified
Source background ring (s) (e.g. number, defining demographics, relevant experience, *applicable bias*)	– Multiple (6) – Recent residents in University of Southampton accommodation. – Families with young children, recently arrived to the UK from countries with hot climates) – Male and female *A. Representativeness* *B. Availability of scenarios*	A. Unusual user group allows mental model descriptions to be attributed to interaction with specified devices, rather than habit (specific mental model descriptions cannot be generalized to native UK population, however). B. Questions regarding background of participant and previous experience with central heating, added to interview.

(Continued)

TABLE 4.1 (*Continued*)

Risk of Bias and Mitigation Strategy for Method Adopted, Derived Using Tree-Ring Method from Revell and Stanton (2012) as Described in Chapter 2

Perspective of Gathering Data	Mental Models in Home-Heating Control	
Source social ring(s) (e.g. number, method of communication, incentive, *applicable bias*)	– Multiple (6) – Answers to structured interview based on Kempton (1986) and Payne (1991) – Concept map representing users mental model description *A. Belief bias* *B. Consistency bias* *C. Embarrassment of incorrect answer* *D. Reluctance to offer 'ridiculous' answer* *E. Misunderstandings – not native English speakers*	A. Topic of study – method seeks user's beliefs of home-heating system function. B. Positioning provided in interview that inconsistencies are 'ok' and expected, to reassure participants. C. To prevent embarrassment, careful positioning reiterated throughout interview that technical accuracy is not sought, only how participant 'imagines' what is happening. D. To encourage free discourse, participants told upfront of the opportunity to verify which parts of concept map they felt 'sure' about, and which they were uncertain of. E. Misunderstandings reduced by requiring subjects to have good levels of spoken English. Paraphrasing by the analyst was used throughout, providing opportunities to check understanding.
Analyst social ring(s) (e.g. number, access method, *applicable bias*)	– Single – Semi-structured interview – Concept map activity *A. Order bias* *B. Belief bias* *C. Experimenter bias* *D. Confirmation bias* *E. Leading by experimenter* *F. Cueing*	A. Order bias acknowledged. B. Analysis table developed to encourage objective categorization to mitigate for belief bias. C. Experimenter bias minimized by conducting interviews 3 months after the study initiated, and taking care not to provide subjects with information about their heating devices that may alter their thinking/behaviour. D. Confirmation bias minimized by analyst avoiding exposure to how the devices in the study function, ensuring participant is required to fully explain their own thinking E. Leading minimized by analyst by providing semi-structured interview template, and piloting interviews prior to data collection. F. Cueing minimized by using participant-initiated terminology and using plain paper (rather than a template) to construct the concept map.

(*Continued*)

TABLE 4.1 *(Continued)*

Risk of Bias and Mitigation Strategy for Method Adopted, Derived Using Tree-Ring Method from Revell and Stanton (2012) as Described in Chapter 2

Perspective of Gathering Data		Mental Models in Home-Heating Control
Analyst background ring (e.g. number, defining demographics, relevant experience, analysis method, *applicable bias*)	– Single analyst – Psychology – Design – British – Compare user mental models descriptions and transcripts to Kempton's (1986) shared theories of home heating *A. Belief bias* *B. Confirmation bias* *C. Bias in interpretation of outputs*	A. To confront belief bias, analyst's own bias accessed by answering the interview questions prior to commencing interviews. B. To aid objective categorization of individual mental models with an analysis table based on Kempton (1986, 1987) was constructed and alternate interpretations to Kempton (1986) were actively sought. C. To minimize bias in interpretation, the mental model diagram (that formed the basis for analysis) was constructed with, and verified by, the participant.
Analyst social ring (e.g. number, *applicable bias*)	Organization of textual and diagrammatic information for Academic Community interested in climate change, behaviour change, device design, psychology, human factors *A. Bias – graphical and descriptive constraints of journal paper*	A. Bias acknowledged.
Analyst cognitive artefact ring (e.g. type(s), *applicable bias*)	One-word analogous textual description. Associated 'concept map' shows key components, and their relationship in terms of links, cause and effect. *A. Bias – single view of mental model*	A. Specific view of mental model chosen to inform pragmatic applicable to design/ instruction.

three main parts: (1) background of participant, (2) self-report of user behaviour and (3) mental model of device function. For a full explanation of the development and examples of outputs and interview template, please see Chapter 3. In brief, part (1) covers the participant's previous experience with home heating devices, determines if they have any formal training that may inform an 'expert' understanding the system and captures the terminology they use when discussing the heating system in their home. Part (2) uses standard questions, probes and scenarios to help build up a

FIGURE 4.2 Participant A's output from the paper-based activity, which formed the basis for analysis.

diagram of 'typical use' of the participant's heating system during a week. Part (3) uses questions and follow-on probes directed by the interview template, to build up a diagram with the participant. The diagram is built up using post-it notes containing the participant-initiated terminology, as the analyst probes the participant to describe the relationship that exist between the components. Concepts are linked by drawing lines between the post-it notes with a pen (see Figure 4.2 for an example of resulting output). To gain insights into cause and effect, and rules of operation, participants were asked follow-on probes such as 'How does the boiler know when to come on/ off' and 'What would happen if you turned the thermostat to its maximum setting?'. Participant responses to these probes were represented on the diagram using arrows and text (Figure 4.2). Following completion of the diagram, the analyst paraphrased each component, link and rule depicted, so the participant could verify if this represented what they imagined (marked with a smiley) or if they were uncertain this reflected what they really imagined (marked with a question mark). Participants were given the opportunity throughout the interview to amend the diagram to better reflect what they thought. These participant-verified diagrams from part (3) of the method represent the device topology and causal model of device function, which De Kleer and Brown (1983) consider fundamental to mechanistic mental models. The form of the verified description is similar to a 'concept map'. They differ from simple concept maps as they contain written descriptions of rules and variables to enhance understanding of the user's 'causal model'. This output is considered to reflect the device model held by the user of their existing home-heating system. Wilson and Rutherford (1989) made the point that the outputs analysts capture when seeking a mental model is distinct from their actual mental model (an internal construct).

Throughout this chapter, the participant-verified diagram is referred to as their mental model *description* of home heating.

4.2.4   ANALYSIS OF OUTPUTS

The user-verified mental model descriptions, and further evidence from the interview transcripts, formed the basis of discussion to determine aim 1: 'if distinct mental model descriptions of thermostat function, which differ significantly from the actual functioning of UK heating systems, can be identified with present-day UK heating systems'. To determine if the shared theories described by Kempton (1986, 1987) and others are evident in the case study group, the interview transcripts were also examined. As the approach was designed with reference to the style and content of questions described by Kempton (1986, 1987), the interview transcripts from the inter-group case study could be examined to find evidence of responses that met Kempton's (1986) criteria of either a feedback or valve shared theory.

To categorize data from the interview transcripts systematically, according to Kempton's shared theories, a reference table was developed (see Table 4.2). Kempton (1986, 1987), provided extended descriptions of valve and feedback model types. These descriptions were content-analysed, identifying four distinguishing themes: (1) user behaviour, (2) thermodynamics, (3) cause and effect and (4) sensing/control. For each of these themes, sub-themes with examples of participant responses considered (according to Kempton 1986, 1987) evidence of a particular theory are shown in Table 4.2. It was then possible to systematically compare the meaning of participant responses in the intergroup case study to those in Table 4.2, to determine if evidence of Kempton's (1986, 1987) shared theories could be found. The following process was adopted when analysing the transcripts, to aid reliability:

1. Examining sections 2, 3 and 4 of the interview transcript, the types of responses by the participant were coded to separate 'meaningful responses by the participant' (the pool for analysis) from 'confirmation of interviewers paraphrase'.
2. Each response in the pool for analysis was coded by the main themes in Table 4.2. An 'other' category was used for responses that fell outside of these themes.
3. Using Table 4.2, each response was evaluated as evidence of Kempton's (1986) shared theories. Whilst Norman (2002) and Peffer et al. (2011) did not provide sufficient descriptions to produce entries in Table 4.2 for 'timer' and 'switch' models, it was also considered if these models could be inferred from participant responses. Please also see Chapter 3, where a reference table for quick analysis of diagram outputs from the method adopted contains inferred responses for 'timer' and 'switch' models. Responses were assigned to the following categories: (1) feedback, (2) valve, (3) timer, (4) switch, (5) ambiguous and (6) N/A.

TABLE 4.2

Analysis Table for Categorizing Responses from Interview Transcripts

Themes		Typical Responses from a Thermostat User Who Holds a Feedback Shared Theory (Kempton 1986)	Typical Responses from a Thermostat User Who Holds a Valve Shared Theory (Kempton 1986)
4. Sensing/control	Sensing temperature	• Thermometer (in thermostat) senses temperature	• Human senses comfort
	Locus of control	• System regulates	• Human regulates – balances heat generated with heat lost
3. Cause and effect	Thermostat set point – cause and effect	• Determines the 'on-off' temperature for the furnace – furnace runs at a constant speed	• Linear relationship – controls the amount or heat/rate of flow – furnace varies in its rate of flow/amount of heat
	Consequence of different set points	• Small increase in set point – furnace on for short time, then turn off	• Small increase in setting – small increase in rate of flow/amount of heat
		• Large increase in set point – furnace run for longer to reach the set point temperature	• Large Increase in setting – large increase in rate of flow/amount of heat
		• Static set point – temperature maintained by furnace switching on/off	• Decrease in setting – decrease in rate of flow/amount of heat
			Implication: confusion may occur if boiler turns off itself
2. Thermodynamics (device within a broader system)	External temperature.	• No impact, as thermostat will sense inside temp and respond to maintain temperature	• When weather is cold, house gets colder, so user compensates by regulating the thermostat
		Implication: confusion may occur if house seems colder on colder days, or if comfort levels are low despite usual set temperature	
	Indoor temperature	• The effect of the house cooling down means more fuel is consumed when you increase the heat again, than if you had left it at the same setting	• Impact of cold bodies not considered to have an effect on energy used – more fuel is consumed at higher settings than at lower ones, as the rate of flow/temperature is higher
		Implication: No night set back	*Implication: Night set back*

(Continued)

TABLE 4.2 *(Continued)*

Analysis Table for Categorizing Responses from Interview Transcripts

Themes		Typical Responses from a Thermostat User Who Holds a Feedback Shared Theory (Kempton 1986)	Typical Responses from a Thermostat User Who Holds a Valve Shared Theory (Kempton 1986)
1. User behaviour	Pattern of adjustment	• Minimal adjustments over short period of time • Will change thermostat set point deliberately at times when different levels of comfort are required; e.g. different levels of occupancy or activity	• Thermostat often changed between each hourly datum • Thermostat is adjusted at least hourly whenever someone is in the house

4.3   CASE STUDIES

Detailed individual case studies are presented of participants A, B and C. These participants were chosen as their mental model descriptions differed significantly from the actual function of the heating system in their homes, as well as from the descriptions of each other. The remaining three participants from the pool of six, provided mental model descriptions with conceptual entities and their interrelationships broadly aligned to the actual functioning of the heating system, with a 'feedback' shared theory for the functioning of the thermostat. As the way users significantly misunderstand the functioning of the heating system is of interest in this study, the results from these participants will not be discussed in detail in this chapter. Each case study will be presented with a description of the participant and their user-verified mental model, redrawn for clarity. Discussion within each case study will focus on the consequence of differences in the content (in terms of elements/conceptual entities) and relationship (in terms of conduits between elements and rules of operation) of these mental model descriptions. How these differences influence user behaviour, and consequences in terms of energy consumption, will also be discussed.

4.3.1   PARTICIPANT A: A FEEDBACK MENTAL MODEL OF THERMOSTAT WITH ELEMENTS OF VALVE BEHAVIOUR

Participant A was a female homekeeper in her thirties. Originally from Malaysia, she lived with her student husband and two school-age children. She and her husband both control heating in their home. Participant A's user-verified mental model description of home heating is shown in Figure 4.3. This has been redrawn for clarity

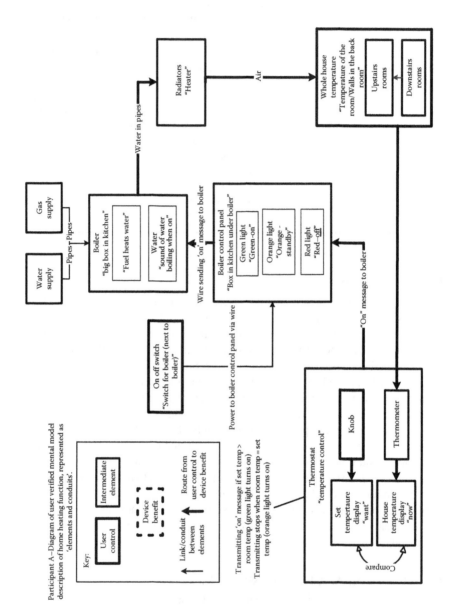

FIGURE 4.3 User-verified mental model description of home heating for participant A.

and the different elements (in rectangles) have been coded according to the key, to distinguish user controls, the device benefit and intermediary elements (between user control and device benefit). The arrows between the elements represent 'conduits' of either information or heat distribution. The thick line shows an example causal route between a user control and the device benefit.

Looking at Figure 4.3, the elements and conduits form a 'loop' shape, which immediately suggests a feedback model. The control devices include the thermostat knob and the main on/off switch for the heating system. When prompted, participant A was not aware of the existence of either the programmer device, or the thermostatic radiators valves. Participant A described her goal when using the heating system was to control the temperature of the whole house.

The second aim of this chapter was to see if distinct mental models of home-heating thermostats could be categorized according to Kempton's (1986) shared theories. To make analysis easier, the elements and conduits which form the causal path for the thermostat have been redrawn in isolation in Figure 4.4. The left-hand element is that which is controlled by the user, depicted with a 'hand'. Automation is represented with a 'robot head', and the rule for automation shown in a thought bubble. Other elements included in the causal path are shown with a square, and the benefit of using the home-heating system is represented by a star shape.

The causal path shown in Figure 4.4 was further amended using evidence from the transcripts. The type of function to be performed by each element is indicated with an icon representing either a discrete or variable function, to give the reader a 'snapshot' of the relationship between elements. The relationship between these functions in specific scenarios provides insights into cause and effect (deKleer and Brown 1983). An insight into the likely behaviour patterns of participant A is gained by understanding how the relationship between these functions allows her to make predictions about the results of her actions.

Looking at Figure 4.4, we can see that participant A's causal relationship includes a thermometer to sense the house temperature and that this feeds back data to the thermostat. The boiler is thought of as functioning like an automatic on/off switch, based on a rule whereby a comparison is made between the temperatures of the set point and house temperature. Referring to Table 4.2, considering sensing temperature, locus of control and thermostat set point, the data from participant A clearly fall into the typical responses provided by a user holding a feedback model type. When considering the consequence of different set points, participant A did not explicitly state that the furnace ran at a constant speed, but this is inferred since she only refers to discrete states of the boiler (on/off/standby) rather than variations in temperature or intensity. The boiler control (which is thought to be the programmer device) provides feedback on this status by different coloured lights. The following quote emphasizes that the boiler control is significant to her model of the boiler regulating the house temperature. When reading the following transcript, please note that participant A initially favoured the term 'thermometer' for the entire thermostat unit. Square brackets in the transcriptions show the intended meaning.

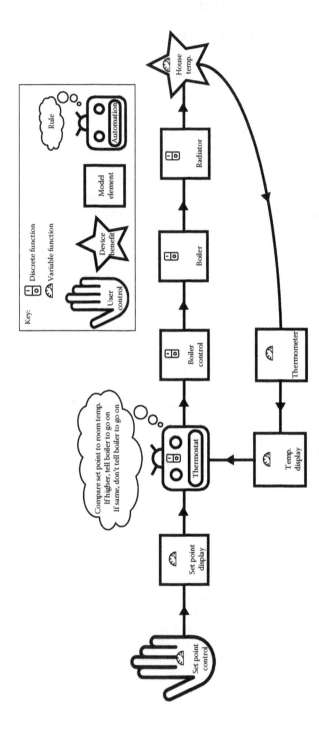

FIGURE 4.4 Isolated causal relationship for thermostat knob taken from participant A's user-verified mental model description of home heating.

Analyst: ....All right, so we've got this idea when you turn the thermom-
                eter [thermostat] up, it [programmer] goes green, normally it's on orange,
                which is our middle one. How does this going green affect what's hap-
                pening in there? ....
Participant A: When the temperature in the house probably down a bit than the
                temperature that we set, that's why it come green.
Analyst: So you're saying you've got the temperature that you set, which is from
                the thermometer [thermostat].
Participant A: Yes, set and then maybe the temperature in the house getting colder
                a bit because of the outside maybe and then .....automatically turn to green
                to make the house warm again to the set temperature.

Participant A is also very clear about when the boiler automatically switches off, stating 'It's nice and warm to the right temperature, it [the boiler] don't do anything' and
'It [thermostat] will tell [the boiler] stop working!'.

The first quote also considers thermodynamics relating to external temperature,
showing again that the device, rather than the user, compensates for changes in external temperature. According to Table 4.2, this is a typical response from a user with a
feedback shared theory.

In terms of patterns of adjustment (Table 4.2), participant A initially provides
clear evidence of a feedback shared theory. She reports adjusting the thermostat to
deliberate set points corresponding to specific times relating to occupancy and activity, as we can see in the following:

Analyst: And can you... do you happen to know what temperature you turn it up
                and down to?
Participant A: At the evening, around 6 or 7, I tend to put it up to 23 then we go
                to sleep we keep it to 21. And, during the day, we normally keep it to 20.

Analyst: So, at sleep it's 21, when you wake up in the morning, do you change it or
                do you just leave it until your children come home?
Participant A: Well, usually I keep it to 20 because I want to keep the house warm
                because my kids go to school and then I keep it 20 because there is sun
                outside and if I'm in the lounge, it gets too hot.

There is an indication, however, that in certain conditions, in certain parts of the
house, the device is not able to regulate house temperature sufficiently to maintain
comfort, which affects participant A's set point choice. This is a response expected
from a user with a valve shared theory (see Table 4.2, external temperature). The
thermostat used is positioned in the hall (see Figure 4.1) in participant A's house,
and actually takes air temperature readings at its location, influenced primarily by
the heat output of the hall radiator. The following transcript reveals that participant
A imagines the inside temperature display represents the temperature of the 'whole
house', rather than a single area.

Analyst: And how does it know what the house temperature is?

Participant A: Usually, the house temperature is... you're not touching anything by the small box you're talking about. If you just look at that, I think that is temperature of the room... of the house.

Analyst: Of the house. So when you say the house, do you mean the walls, or...?

Participant A: The whole house ..

This misinterpretation in her mental model is significant in terms of her self-report of behaviour patterns. Whilst originally describing a regular routine for set point change, when provided with a typical scenario where comfort levels are too high, she revealed that ad hoc adjustments also occurred.

Analyst: Next one scenario 3: The heating is on, you can feel the radiators are on but you've been rushing around doing housework or exercise or looking after the children or cooking and now you feel really hot and uncomfortable, what would you do?

Participant A: I would scream my husband put it off!

Analyst: And when you say put it down, you mean this button [point to thermostat].

Participant A: Yes, I tell my husband.

Analyst: O.K. your husband.

Participant A: Because I do exercise at home, I work out at home, I do my exercise at home.

Analyst: How typical is this and how often would that happen?

Participant A: Er, when I go to sleep, actually, every night. Because sometimes he has a habit to make it to 23. I think because he's sitting all the time. When I go upstairs to go to sleep, I scream you've got it so hot!

Further exploration revealed that this 'comfort battle' between her husband and herself occurs regularly during the day. If her husband is home, he frequently increases the thermostat set point to 22°C, and she returns it to 20°C. This pattern of behaviour is closer to that expected from a user who holds a valve shared theory (see Table 4.1). Kempton (1986) stated that conflict battles may result in valve behaviour patterns. However, participant A reports that the battle occurs when she and her husband are located in different parts of their home, where thermodynamics of the house structure also affect comfort levels. As participant A believes the 'whole house temperature' is measured by the thermostat, it is logical, when considering the causal path in her mental model description (Figure 4.4), that comfort would be regained by adjusting the set point, rather than a different strategy. More appropriate strategies, such as adjusting the thermostatic radiator valves to different settings to accommodate both her husband (studying downstairs) and herself (exercising, or sleeping upstairs) cannot be considered, as TRVs are absent from her mental model description of home heating (Figure 4.1). As a result, heat energy is wasted by overheating rooms where a facility to limit the temperature to comfortable levels exists.

Participant A clearly provided a mental model description of thermostat function that fit the criteria derived from Kempton (1986) of a 'feedback' shared theory. Self-reported planned, regular set point adjustments also fit the behaviour expected from a user holding this theory. Due to comfort conflicts, participant A reported behaviour patterns which, if viewed from set point measurements alone, would indicate a valve theory was held. However, it is argued that this behaviour pattern is driven not just by 'comfort conflict', but by misinterpretations in the thermostat mental model, as well as an incomplete mental model description for the heating system as a whole.

4.3.2   PARTICIPANT B: FEEDBACK BEHAVIOUR WITHOUT A FEEDBACK MENTAL MODEL

Participant B is a female student in her late 20s. Originally from Mexico, she lives with her husband and young child. Participant B's mental model description for home heating is shown in Figure 4.5. She is the sole operator of heating in her home.

We can see from Figure 4.5 that the mental model description of home heating developed with participant B is more complex and detailed than that for participant A. The shape contains no 'feedback' loop input to the thermostat device, which suggests the user does not hold a feedback shared theory of home heating. Participant B describes multiple control devices, including the program schedule, the override button, the thermostat set point control, boiler control knob and radiator control knobs (TRVs) for each radiator. The only heating control absent from this description is the power switch for the whole heating system. Participant B similarly described that the purpose of using the heating system was to warm the whole house. Figure 4.6 shows the causal path of the thermostat control, isolated from the other control devices, with icons added to represent the function of each element as inferred from the interview transcript.

From Figure 4.6 we can see that participant B's causal relationship for the thermostat set point is dependent on the water supply. There is no thermometer or feedback loop included in the mental model description by participant B. Variations in set point are matched with variations in the volume of water to be heated up by the boiler. In general usage, the boiler operates according to discrete states, automatically switching off when the corresponding amount of water has been heated. A larger volume of heated water (following, according to participant B's model, from a higher set point value), takes a longer time period to run through the radiators, resulting in a greater amount of radiated heat, and ensuring a higher house temperature.

When comparing Figure 4.6 to the criteria in Table 4.2, the lack of thermometer refutes a feedback theory being held. However, when responding to questions regarding the boiler override control, it appears that participant B is aware there is an inside temperature display on the thermostat device (see the following transcript). She did not consider this of significance, however, when producing her mental model description of home heating (Figure 4.5).

Analyst: ...... what situations do you normally go in and press the button to get it to come on?....

Participant B: Basically if I see the temperature is 16 or 17 degrees I feel it's a bit cold.

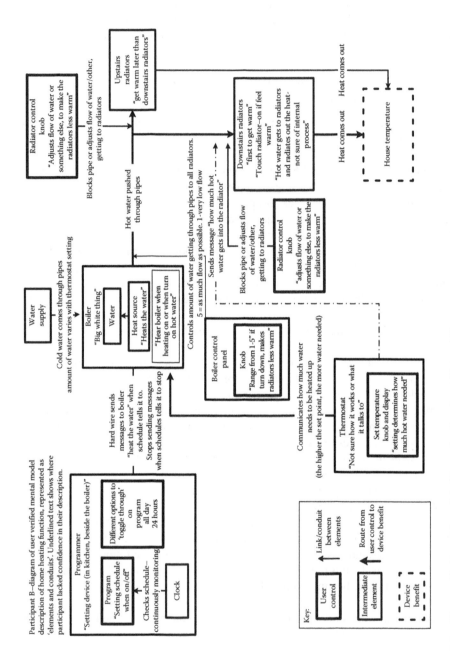

FIGURE 4.5   User-verified mental model of home heating for participant B.

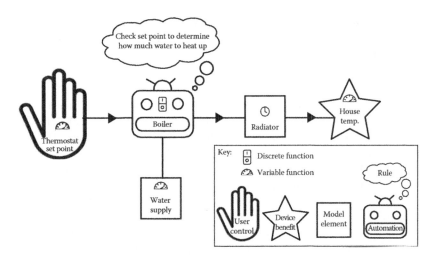

FIGURE 4.6 Isolated causal relationship for thermostat set point taken from participant B's user-verified mental model description of home heating.

Analyst: Okay. So even if you perhaps didn't feel cold but you noticed it was cold then perhaps turn it on, or is it only if you feel cold?

Participant B: Yeah, I feel cold and then I check the temperature and say, "Okay, I'll turn –

Analyst: So you making sure you really are cold?

Participant B: Yeah, that it's not just me. (Laughs)

Analyst: If you're cold. So you check... and that's on the thermostat?

Participant B: Thermostat.

This transcript clearly shows that participant B uses the device to determine if the house temperature is appropriate, rather than her own sense of comfort, which from Table 4.2 is characteristic of a holder of a feedback, rather than valve, mental model. Considering cause and effect (Figure 4.6), there is a clear linear relationship between the thermostat set point and the amount of water heated, suggesting a variation on the criteria for valve theory (Table 4.2). We could consider participant B has a 'valve' theory, for the relationship between the thermostat with the water supply, but not with the boiler intensity.

Analyst: .... So what would happen then if you had the thermostat, like, up to 30 or something, if you had the thermostat at a really high temperature, the maximum temperature that you can go to, what do you think would happen?

Participant B: Well then the boiler will have to produce more hot water and keep going and going through the radiators until they reach the temperature, I mean they make the temperature to 30.

Analyst: So if 30 degrees is set on the thermostat lots of hot water is made Okay. Alright. What would you think would happen if you turned it right down to 5 degrees?
Participant B: The boiler will not operate

A more appropriate analogy may be that each set point temperature has a fixed volume of water associated with it. When adjusting the set point, a message is sent to the boiler to select and heat up the corresponding volume of water. This also is clearly different from a feedback theory, whereby the boiler operates until the house achieves set point temperature, sensed by a thermometer. However, it may lead to similar behaviour, as it is accepted by participant B that the system ensures through this selection of water volume that the desired house temperature will have been reached when the boiler deactivates. Continuous adjustments of thermostat set point would therefore be unnecessary. The consequence of different set points, when considering the relationship with water volume, again can be interpreted as belonging to 'valve theory', as larger amounts of heated water result from larger increases in set point.

The preceding transcript, however, could also be interpreted as 'feedback theory' according to Table 4.2, despite following from a mental model description that lacks a feedback loop. Heating larger volumes of water, assuming the boiler runs at a constant speed, also predicts the furnace will run for longer periods. When questioned, participant B was clear that the water temperature in the boiler was not raised by the thermostat, so a valve theory for boiler operation, as described by Kempton (1986), is refuted.

When questioned about her behaviour patterns when operating the thermostat, participant B reported keeping the set point at a single value, relying on the programmer to regulate the heat. This behaviour again fits better with a response expected from a holder of a feedback, rather than valve, model (Table 4.2).

Analyst: ....Okay, so you're saying the thermostat's normally set at 20, sometimes you notice it's cold, would you then when you notice it's cold turn the thermostat up or just go to the programmer and press the programme?
Participant B: Go to the programmer.
Analyst: Okay. .... is there any reason you would go and change through your thermostat?
Participant B: No.
Analyst: No? Okay. So not during the week or weekends or holidays or anything like that, you would just keep it how it is and make everything controlled through the programmer?
Participant B: Yeah.

A goal of this chapter was to see if distinct behaviour patterns of thermostat function could be categorized according to Kempton's (1986) shared theories. This has been difficult to achieve with participant B's mental model description. Many traits which would be expected from a user holding a feedback theory were discovered. However, the concept of feedback, or temperature sensing was not present in her verified mental model description (Figure 4.5). The linear relationship between the thermostat set point and the water supply was more akin to a valve theory; however, it should be noted that this is not with the sense intended by Kempton (1986), which relates to heat flow or intensity in the boiler. Kempton (1986) suggested that users with a feedback shared theory could unnecessarily waste energy by avoiding 'night set back' of the thermostat. Participant B uses the programmer to limit consumption at night, so the need for night set back is negated regardless of the theory held. The reasoning of feedback theory holders, according to Kempton (1986) (and summarized in Table 4.1) is that it will take more energy to reheat a cold house than to maintain a warm house. This thinking is evident with participant B in the following quote. From participant B's transcript, neither the thermostat nor the programmer settings are adjusted when away from the house at the weekend, or on holidays, suggesting energy is wasted by heating an unoccupied house.

Participant B: Well I was actually thinking with that setting that I need... or that the house needs to be heated for not such a short time because maybe it will have to work more the next day. I mean to not leave the house unheated for a long period of time.

The desire to identify distinct mental models of device function comes from the benefits of being able to predict user behaviour. Participant B, whilst using a deliberate thermostat set point, does not habitually make adjustments (preferring to use the programmer and override device). Figure 4.5 shows uncertainty in much of the mental model description relating to the thermostat device, with underlined areas showing low confidence in the description. Participant B's mental model for thermostat function may be poorly developed because she does not use the thermostat. On the other hand, participant B may not use the thermostat because her mental model of the thermostat is ambiguous or incomplete, undermining confidence in the outcome of her actions. The ability to predict behaviour patterns relating to user mental models of device function is necessary if design strategies to improve usability or influence behaviour are to be successful. This association is clearly more likely to be found when the participant actively uses the device under investigation.

4.3.3   PARTICIPANT C: TIMER MODEL FOR ALTERNATE CONTROL DEVICES

Participant C was a female student in her 20s. Originally from Brunei, she lived with her husband, young child and father-in-law, and is the sole operator of the heating system. Figure 4.7 shows her user-verified mental model description of home heating.

Participant C's mental model description of home heating is linear and branched in shape, and shows three different user controls: the schedule for the timer, the on/off (or override) switch for the programmer and the radiator knob (TRV). The device benefit is not considered the house temperature, as with participants A and B, but rather her own thermal comfort. Conspicuous by its absence is the thermostat device.

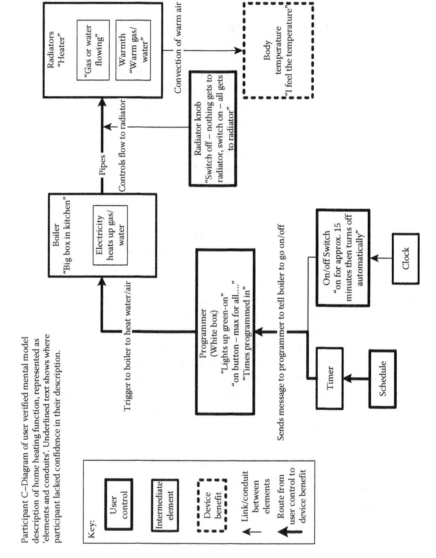

FIGURE 4.7 User-verified mental model description of home-heating function, from participant C.

The second aim of this chapter was to determine if mental model descriptions of thermostat function can be categorized according to the shared theory types in the literature. It is clear from the lack of thermostat device in Figure 4.7 that this type of analysis couldn't be undertaken. To further the initial aim, that distinct mental model descriptions of home heating exist, two causal paths will be described from participant C of alternate control devices, as they provide further insights into the consequence of incomplete mental models. The causal paths of the program schedule and of the boiler override button are shown in Figures 4.8 and 4.9. Unlike the previous causal paths, which focused on the thermostat, these include icons representing a time-based variable.

Participant C clearly describes the programmer unit as a time-based automatic switch responsible for turning the radiators on and off:

> It is a digital device for the system to say at what time you want it to come on, and it will light up to yellow colour if the radiator is on and it will light up to red if the radiator is off, and it is automatically turned to on and off by the program.

Figure 4.8 displays a linear route from manual input of the schedule to the resulting increase in body temperature, with the schedule function being time-based. The timer button (program option on the programmer), boiler and radiators are reported in the transcripts as being either 'on' or 'off' and therefore function according to discrete, rather than variable, states. This is an appropriate mental model for the programmer, if used in isolation. It is not clear from the transcript if the benefit, in terms of body temperature, is directly linked to the schedule, so an icon is absent from this element in Figure 4.8.

Participant C reported that when using the override switch, the boiler switched itself off after approximately 15 minutes. The following transcript describes her confusion about this outcome. The blank lines represent parts of the conversation that have been omitted as they do not add new insights.

Analyst: How do you think it knows when to come off? .... is it from the timer? Or something else that makes it switch off?
Participant C: It can't be the timer because the timer is already being programmed so it must be something else. It's by default it will just go off.

Analyst: So maybe after a certain period of time, or?
Participant C: Yeah.
Analyst: Okay. So you've kind of got this idea about whenever you switch it on, it's almost like it's only on for a short period.
Participant C: Yes it doesn't obey me.

From the last comment it is clear that participant C does not feel in control of the effects of this device. Further discussion led to the idea of a 'clock' telling the program when to switch off. The uncertain language used in the preceding transcript ('it must be', 'it can't be') means it is likely that this link was a construct of the interview, rather than the participants' existing mental model of the user (Payne 1991). It is

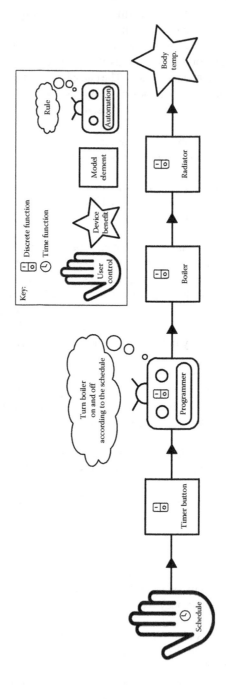

FIGURE 4.8 Cause and effect route of schedule function for participant C.

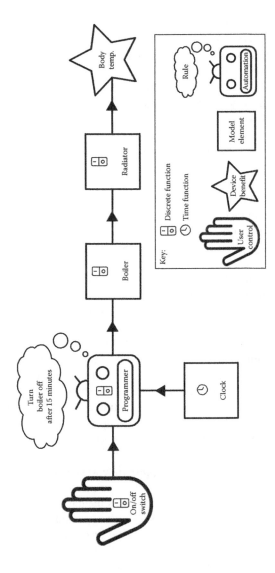

FIGURE 4.9   Cause and effect depiction of override button for participant C.

argued that the need to add to a mental model to explain an unexpected phenomenon shows that the existing model of the system is not fully functional for this participant. Moray (1990) emphasized how it was not possible to troubleshoot problems successfully if the mental model held by a user lacked the elements responsible for the problem source. It is likely that the true cause of the boiler deactivating a short period after the participant presses the override button is the thermometer registering its surrounding air temperature equal to the thermostat set point (presumably set to a value below participant C's comfort levels). Since a thermostat device and thermometer are both absent from this participant's mental model description (Figure 4.7), this conclusion could not be arrived at by participant C herself.

Participant C relies on a daily schedule, which she believes ensures the boiler operates to produce heat for a short period in the morning, an hour for lunch, and a longer period in the evening. The schedule is repeated 7 days a week, regardless of changes in activity. When asked when she used the override button, participant C revealed that she regularly overrides the system when the heating is scheduled to be on (between 5 and 10 pm). Participant C does not appear to see a conflict with her mental model of the timer, which is turning the boiler 'on and off' according to the schedule, and the need to override the program during scheduled 'on' times. She also admits to using an electric portable heater or 'blower' to maintain or increase comfort levels between, or during scheduled times. These additional actions suggest that the home-heating system with the existing schedule and (unknown) set point, fails to meet the comfort needs of participant C. Depending on how energy-hungry the model of portable heater is, and the frequency and durations of operation, this strategy may result in higher levels of consumption than if the central heating had been set up to provide adequate comfort levels. However, it is possible that this proximal ad hoc heating from an alternate heat source is more energy-efficient at meeting comfort needs than if the whole house was heated for longer periods.

When provided with scenarios where her comfort levels are too low, participant C provided a range of strategies unrelated to her mental model description of home heating (Figure 4.7). These included washing hands, showering, putting on the portable heater, moving away from the window, putting on a blanket and even lighting candles. It is inferred from this response that participant C considers the central heating system in her home to be one of a range of solutions for achieving comfort, rather than the key solution. As her goal for using the heating is body temperature, rather than house temperature, it may be more appropriate to gain this participant's mental model of comfort, rather than device function.

Whilst the lack of thermostat in participant C's mental model description meant it was inappropriate to evaluate if Kempton's (1986) shared theories were present, the transcript was examined for evidence relating to generic 'valve' or 'feedback' concepts. When participant C was asked how she knew the heating was on, she reported that 'I will... first thing I will think house is warm',suggesting that she is sensing comfort, rather than referring to a device to get feedback of boiler activation. When describing how she thought thermostatic radiator valves work, she similarly described valve thinking 'The amount of the flow of the heating material flowing through the radiator. So if this is being turned on then there'll be more flowing into the radiator. There is no sense of variations in temperature or intensity of heat as a result of the heating system, however. Feedback concepts were completely absent from the transcript. For this participant, further exploration of timer mental models of home heating would be appropriate.

4.4 DISCUSSION

A small-scale inter-group case study was undertaken to identify: (1) if distinct mental models of the way home-heating systems function could be found that differ significantly from the actual functioning of UK systems, and (2) if these models could be categorized according to 'shared theories' of thermostat function from the literature. Data were collected from six participants in matched environments using a semi-structured interview including paper-based activities, (see Chapter 3 for further details). The impetus for the research was to understand if the mental models held by individuals could explain their energy-consuming behaviour in a way that could usefully inform strategies aimed at reducing consumption through behaviour change.

It was found that three of the six participants produced mental model descriptions of their home-heating system that differed significantly from the actual functioning of their system. The differences explained their self-reported behaviour with their home-heating system, which, by association, may also explain their levels of energy consumption. From the mental model descriptions produced by the six participants, four represented thermostat function in a way that could be categorized according to Kempton's (1986) 'feedback' shared theory (a simplified version of the actual home-heating functioning). One participant described a variation of Kempton's (1986) 'valve' theory, and another participant omitted the thermostat device from her description, preventing comparable categorization. Timer and switch shared theories for the thermostat were found useful for categorizing user mental models of alternate home-heating controls, such as the programmer and boiler override button. The valve shared theory could also be applied in a general sense to user mental models of the TRVs.

The key findings of this study that add to the existing body of knowledge are that (1) UMMs of thermostat function can be found that fall outside of the feedback, valve, timer and switch shared theories described by Kempton (1986), Norman (2002) and Peffer et al. (2011), and (2) omissions of entire control devices from UMMs of home heating were evident despite an environment matched in terms of dwelling, and type and layout of home-heating devices. In addition (though not emphasized in the reporting of the three in-depth case studies), thermostat set point adjustment was less prevalent than expected, with four out of the six participants reporting a reliance on other devices (e.g. programmer, override button and TRVs) when adjusting their home-heating output. Finally, the discrepancy of the user's goal when using the heating system (e.g. heating the whole house or increasing the comfort of the occupant) and the actual benefit of the system (i.e. to produce a constant rate and intensity of heat, limited in period of operation by the program schedule, in activation by the thermostat set point and in output by the TRVs set point) was also helpful when explaining reported confusion by the user, when operating the system.

In itself, the existence of UMMs that cannot be easily categorized according to existing shared theories is only important in terms of energy consumption if the associated behaviour is significantly more or less 'wasteful'. In this study, the associated (self-reported) behaviour of participant B closely matched that expected of users holding a 'feedback' theory, so participant B's specific UMM is not of special interest. However, consensus in the literature (e.g. Kempton 1986, Norman 2002, Peffer et al. 2011, Richardson and Ball 2009) that UMMs fall into these categories of shared theory is challenged. This suggests that the existence of additional 'shared theories' or unique individual UMMs, which may have a significant impact in terms of energy-consuming behaviour, cannot be ruled out.

The omission of control devices, rather than merely settings/options on a device, is an important finding. The omitted devices in our study included key controls, such as the programmer and thermostat, which hold a significant role in allowing the user to optimize energy consumption. The lack of key controls in UMMs affects the strategies that users can adopt in order to meet their goals. For example, it could impede users from achieving their desired level of comfort by being unaware they could adjust the thermostat or TRV setting, or that they could set the programmer to start heating before getting up. Alternatively, it could prevent users from reducing consumption by being unaware they could reduce the thermostat or TRV setting, or that they could set the programmer to automatically switch off at times when the home is typically unoccupied. Failure to meet user goals, as in the case of participant C, who reverted to an electric heater when failing to meeting her comfort goals, could encourage alternate strategies that may be more costly in terms of energy consumption than optimal operation of the home-heating system. In addition, the lack of key controls in UMMs of their own heating system could hinder the success of advice-based strategies to encourage reduced consumption in the home, when based on changes to the way users operate 'omitted' devices. For example, government advice in the United Kingdom to 'turn down your thermostat by 1 degree' (www.energysavingtrust.org. uk) would have little effect on participant C in this study, who omitted the thermostat control when producing her mental model description of home heating.

Whilst the numbers in this inter-group case study are too small to indicate a trend, the preference of the majority of participants in this study to favour the programmer and boiler override devices to the thermostat, when asked about ad hoc or routine adjustments, was surprising given an emphasis in the literature on thermostat behaviour styles (e.g. Kempton 1986, Norman 2002, Richardson and Ball 2009, Peffer et al. 2011). If large proportions of the general UK population also seek out alternate control devices to the thermostat, when making adjustments, it calls into question the present-day relevance of Kempton's (1986) insight. It may no longer be important to consider users' 'shared theories' of thermostat function and associated behaviour patterns as a means of understanding domestic energy consumption. However, this inter-group case study does suggest that a link between UMMs of home heating and their strategies for controlling heating, so applying Kempton's (1986) insight at the system level, incorporating the integration of a range of control devices, rather than to a single control device, may be more appropriate.

When a user needs to translate their home-heating goals (e.g. comfort, reduced consumption) in terms of the options available on the home-heating system (e.g.

home-heating control set points, options and schedule durations), the ease of this translation is likely to affect optimal operation. This link between goals, mental models, strategies and performance is clear in the literature for interaction with devices (Norman 1986) and within complex systems (Moray 1990, Bainbridge 1992). In this case study, not only was there evidence of users not being able to meet their goals, with participants A and C, but also a misunderstanding of the benefit that the home-heating system could provide (e.g. participants A and B, who thought the thermostat setting ensured the whole house was maintained at the chosen temperature). A better understanding of how to encourage optimal behaviour with home-heating systems would benefit from considering not only how well the heating system can accommodate user goals, but also how appropriate users' expectations were of the heating system's benefit.

The data collection approach was chosen to achieve key objectives for this inter-group case study: (1) to encourage accuracy in the capture of UMMs by considering bias in interpretation in its development, (2) to capture a description of the user's device model, (3) to represent the user's device model description in a concrete diagram form, to aid design-based strategies, (4) to produce data that allow categorization of UMMs of thermostat function by existing shared theories in the literature, (5) to produce data that allow alternate theories to be considered.

The method developed was intended to enhance previous methods of data collection from established researchers such as Kempton (1986) and Payne (1991) through mitigation of bias and reduction of ambiguity in the outputs. However, a systematic method of considering bias was used in the development of the method, with mitigating strategies or acknowledgement of bias, as stated in Table 4.1. This provides readers with transparency so they can appreciate the steps undertaken to increase the chances that the captured mental model descriptions reflect the user, rather than the analyst. The paper-based element of the method was well suited in structure to capture users' device models of home heating. This concrete representation clearly identified missing elements and misunderstandings, providing emphasis for targeted strategies aimed at enhancing UMMs to encourage energy-consuming behaviour.

In terms of capturing and depicting the user's device model, the process to develop and the resulting form of the user-verified mental model description are considered. Considering the method's ability to categorize UMMs according to shared theories in the literature, Table 4.2 was helpful in categorizing according to Kempton's (1986, 1987) feedback and valve shared theories, but switch/timer shared theories needed to be inferred as Norman (2002) and Peffer et al. (2011) gave insufficient information to populate the categories derived through content analysis of Kempton's descriptions. However, the readers are assured that any evidence that the users held timer and switch theories for thermostat function in the inter-group case study was not overlooked. The concepts of timer/switch theories and Kempton's valve theory are clearly applicable to other home-heating control devices in the study, both appropriately and inappropriately applied. A more generic (e.g. control device independent) definition of these theories would aid applicability to a range of energy-consuming devices and could therefore have broader reach in understanding non-optimal user behaviour. The semi-structured style of the interview was beneficial in allowing participants to freely express ideas that were outside the scope of the shared theories from the literature. This was seen with participant B's variation on Kempton's (1986) valve theory for the thermostat.

When researching mental models, as with any internal construct, there are limitations as it is not possible to directly assess and characterize the concept (Zhang et al. 2010). The method for extracting information and describing the user mental model will always be subject to bias based on the decisions, perspective and requirements of the analyst (Wilson and Rutherford 1989, Bainbridge 1992, Zhang et al. 2010, Revell and Stanton 2012). In Chapter 2, Revell and Stanton (2012) argue that by systematically evaluating and stating the bias in data collection and analysis, the risk to bias for the resulting mental model description is explicit. This particular method necessarily accessed mental model subsequent to device use, due to the nature of the domain. Because of this, there is a risk of models being spontaneously created as part of the process (Payne 1991). The language used and hesitancy of answer did provide an indication of this effect. However, the authors believe this is an unavoidable consequence of the interview process, which gives an opportunity for the user to reflect upon and refine their mental model. Nevertheless, refinements of a mental model that lacks key elements or contains a misunderstood relationship between components are unlikely to compromise the insights gained by the analyst. The fundamental misunderstandings or omissions, rather than refinements on these understandings, are most illuminating in terms of determining strategies to encourage energy-saving behaviour. There are clear challenges when trying to validate any internal construct, as direct access is not possible. Nevertheless, where there are clear absences of key components of the home-heating system, in the mental model description, or unexpected interpretations of cause and effect, it is unlikely through the choice of probes and construction of diagrams that the participant constructed an alternate view whilst actually holding a more accurate mental model of the heating system.

Another limitation of this research is the general applicability to a typical UK population due to the characteristics of the user group and the small sample size. This group's limited exposure to home heating was, however, a benefit in being able to attribute the UMM descriptions to experiences with the devices present in the matched accommodation, rather than the extensive experience a typical UK resident would have. As the user group was in matched accommodation, it also means the device layout and model types can be regarded as a starting point from which to consider design-based strategies to encourage behaviour change that take into account the misunderstandings and omissions seen.

Further work needs to be done with a larger sample size and more typical UK population to determine if the specific misunderstandings and omissions (in terms of the elements and relationships of components present in users' device models of home-heating systems) that were found in this inter-group case study are more generally applicable. A larger sample would also better indicate the current prevalence of existing shared theories from the literature, applied either to the thermostat or more generically to alternate control devices. It may also reveal new shared theories to add to those already present in the literature. To understand if the device model or shared theory held translates to significant differences in users' energy-consuming behaviour, additional work needs to be done. A study that collects actual user behaviour with the typical range of home-heating control devices, in addition to users' device models of the home-heating system, could shed light on this relationship. There is presently much interest in how to improve consumption with heating systems through design, focusing on programmable thermostats (e.g. Combe et al. 2011, Peffer et al. 2013), as well as investigation into the benefit of support aids for central heating (e.g. Sauer et al. 2009). The central heating system needs to be considered as just that, a system. Sauer et al. (2009) consider home heating to be the most complex system in the domestic domain. Consideration of individual control devices in terms of usability is clearly important, but users need to be aware of which control devices to use in which situations to fulfil their comfort and consumption goals. Support aids for central heating, or the redesign of central heating devices, which promote a functional mental model to users of the integration between components of the heating system, could provide benefits beyond enhanced feedback or proscriptive guidance. It is proposed providing users with a pragmatic understanding of cause and effect for the multiple control devices presented on their heating system that allows them to: (1) relate their actions to their individual goals, and, just as importantly, (2) understand when their individual goals cannot be met by the system. This could not only reduce consumption, but enhance comfort, providing the 'optimal consumption' we all seek.

4.5 CONCLUSIONS

Through comparison of user-verified mental model descriptions of home heating, distinct mental models that differ significantly from the actual functioning of UK heating systems were shown. Evidence of Kempton's (1986) feedback shared theory for the functioning of the thermostat was found. Other shared theories from the literature were useful in categorizing alternate control devices. A user mental model of the thermostat that could not be assigned to the existing shared theories in the literature was also found. Differences in the control devices present in mental model descriptions could explain confusion in operation. Differences in the causal path for specific control devices explained differences in reported behaviour. Variables such as the type and number of controllers and assumed device benefit appeared to interact with mental models of device function to explain reported behaviour. To apply the notion of mental models of home heating to encourage optimal energy consumption misunderstandings and omissions in user mental models could target design focus to encourage users to hold more integrated, functional models. Further work is needed to prove the association between shared theories categorized by this method and actual behaviour with devices, as well as methods for applying mental model descriptions in design strategies.

5 When Energy-Saving Advice Leads to More, Rather Than Less, Consumption

5.1 INTRODUCTION

This chapter uses a combination of data sources to investigate hypotheses 1 and 2 described in Chapter 1. Data relating to mental models of home heating and self-reported behaviour with controls from applying the QuACk method described in Chapter 3, as well as automated data collected from case study households relating to thermostat set point changes, internal temperatures and boiler 'on' periods were compared. These are used to investigate how home-heating goals (e.g. comfort and consumption) are influenced by patterns of device use (hypothesis 2), as well as how mental models of home heating explain the patterns of device use displayed (hypothesis 1). By considering both of these hypotheses within the same case study, this chapter links three out of four concepts depicted in Figure 1.1. This linking of concepts lends considerable support to the validity of the key aim of this chapter, that design strategies can increase home-heating goal achievement by using the notion of mental models to alter behaviour with heating controls.

Domestic energy consumption accounts for approximately 30% of UK consumption, and 60% of this is as a result of space heating (Department of Energy and Climate Change (DECC) 2013). Home-heating devices, therefore, indirectly control approximately 18% of the United Kingdom's entire energy production. In the United States, Peffer et al. (2011) estimate 9% of domestic consumption is controlled by the thermostat device. Their review of thermostat usage, however, revealed almost half of homeowners only sometimes, rarely, or never programmed programmable thermostats, and around 40% of people with manual thermostats did not set back the temperature at night to save energy.

Shipworth et al. (2009) found that average maximum room temperatures, or the duration of operation, were not reduced with the introduction of standard heating controls recommended by DECC to save energy. Increasingly, we find that technology touted to save energy does not deliver on the promises made. The benefits of smart meters have been put into disrepute. High installation costs fail to justify the predicted 2% saving on an average bill. Further, this meagre reduction is conditional on consumers changing their behaviour to actively cut energy use (BBC 2014). Programmable thermostats have been stripped of their Energy Star certification since 2006 in California (DECC 2013), and from 2009 in Washington, DC (Kaplan 2009),

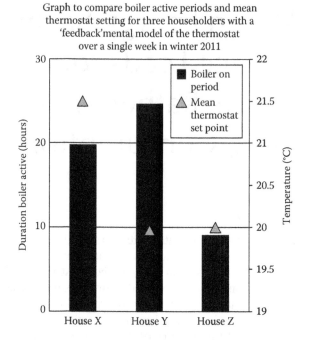

FIGURE 5.1 Graph to compare boiler 'on' periods for three matched households over a single week during winter in the United Kingdom.

as significant savings were wholly dependent on consumers knowing how best to use them. The role users play in operating technology, therefore, cannot be overlooked, and a body of research has found that significant variations in energy use are due to behavioural differences of householders (Lutzenhiser and Bender 2008, Dalla Rosa and Christensen 2011, Fabi et al. 2012, Aerts et al. 2014).

Key causes of variations in domestic heating usage patterns include technologies, habits, knowledge and meanings (Gram-Hanssen 2010). Householders' knowledge of home-heating technology and resulting habits of behaviour are considered in this chapter through the examination of thought processes that guide users' understanding and actions with technology. An important thought process used in this way is termed a 'mental model'. Mental models can be thought of as a 'picture of the world' held in the mind (Veldhuyzen and Stassen 1976, Johnson-Laird 1983, Rasmussen 1983). Whilst householders may not be consciously aware of it, they are using this internal representation to help operate systems, such as their home-heating system (Kempton 1986). This same representation is also used to predict the effects of their actions on, for example, their comfort and consumption levels, and to help understand and explain the changes that occur (Craik 1943, Gentner and Stevens 1983, Kieras and Bovair 1984, Rouse and Morris 1986, Hanisch et al. 1991). The explanations formed may then lead to further actions (Norman 1986). Depending on their experiences, the content of peoples' mental models varies (Norman 1983, Moray 1990, Bainbridge 1992). Consideration of householders' mental models of the home-heating system would therefore be helpful when looking for insights into the cause of variations in people's behaviour with home-heating systems (Kempton 1986, Revell and Stanton 2014). This in turn could focus strategies to realize the potential of 'energy saving' technology.

Kempton (1986) hypothesized a causal association between users' mental models of the way thermostats work, and their behaviour patterns over time when manually changing the set point of the home-heating thermostat. Kempton (1986) identified two typical types of mental models of how the thermostat functioned that were analogous to 'valve' and 'feedback' mechanisms. Those who held a 'valve' model predicted that if the thermostat was turned right up, the house would heat up more quickly (like turning a tap on full to fill up a basin more quickly). Those who held a 'feedback' model predicted that the speed to heat a house was not affected by choosing a higher than desired set point on the thermostat. He found most of his participants had common elements in their individual mental models that fit either one or the other of these types. Kempton (1986) termed these types 'shared theories' rather than mental models. This highlights that they are shared by a particular social group, and represent the common elements of individuals' mental models (Kempton 1986, Revell and Stanton 2012).

There are many definitions of mental models and different perspectives from which to consider them (Wilson and Rutherford 1989, Richardson and Ball 2009, Revell and Stanton 2012), so specificity in definition is key (Bainbridge 1992, Revell and Stanton 2012). The mental model from which Kempton (1986) derived his shared theories, are best understood in terms of a user mental model (Norman 1983) and device model (Keiras and Bovair 1984). That is to say, a mental model held by a user of a specific technology that contains information about the operation and function of that device, and has been accessed and described by an analyst. Kempton (1986, 1987) suggested that distinct behaviour patterns are associated with 'valve' and 'feedback' shared theories of thermostat function. He proposed that characteristics of the feedback shared theory could result in energy being systematically wasted. Since Kempton (1986), no further work has explored in detail the consequences of how referring to a feedback shared theory when operating home-heating controls affects energy consumption.

Revell and Stanton (2014) built on the work of Kempton (1986) and extended this sentiment to the range of home-heating control devices commonly available in present-day homes. However, whilst agreeing that consideration of individual control devices in terms of usability and the mental model held is clearly important, Revell and Stanton (2014) (Chapter 3) argue that users need to be aware of which control devices to use in which situations. This means that the householder needs to be able to adopt an appropriate home-heating 'strategy' to fulfil their comfort and consumption goals. This requires householders to have an appropriate understanding of home heating at a system level, as well as at the device level.

Sauer et al. (2009) consider home heating to be the most complex system in the domestic domain. For complex systems, both Bainbridge (1992) and Moray (1990) describe how the user's mental model of the system constrains the performance with the system. They also emphasize that other variables, such as user goals, influence the resulting strategy adopted with the system. Norman (1986) also highlighted the link between goals, mental models, strategies and behaviour when users interact with individual devices or interfaces. As the strategy adopted by the user ultimately drives the observed behaviour, efforts to change householders' energy-consuming behaviour would benefit from further understanding of the cause and consequence of home-heating strategies.

Following on from Chapter 3, this chapter examines how mental models at the system level affect the strategies adopted at the device level, even when the same feedback 'shared theory' of a home-heating thermostat is held by householders. Further, the impact of differences in the chosen strategy on domestic energy consumption is explored. The 'feedback' shared theory of the thermostat was described by Kempton (1986:80):

> According to the feedback theory, the thermostat turns the furnace on or off according to room temperature. When the room is too cold, the thermostat turns the furnace on. Then, when the room is warm enough, it turns the furnace [Boiler] off. The setting, controlled by a movable dial or lever, determines the on-off temperature. Because the theory posits that the furnace [Boiler] runs at a single constant speed, the thermostat can control the amount of heating only by the length of time the furnace [Boiler] is on. Thus, if the dial is adjusted upward only a little bit, the furnace will run a short time and turn off; if it is adjusted upward a large amount, the furnace must run longer to bring the house to that temperature. Left at one setting, the thermostat will switch the furnace on and off as necessary to maintain approximately that temperature.

This theory is a simplified but essentially correct understanding of how the thermostat works. However, as Kempton (1986) points out, it does not consider the impact on comfort of different levels of infiltration through the home, nor does it include the importance of internal temperature levels on the rate of heat loss. The former point relates to how different parts of the house will heat up at different rates depending on house structures and heat flow between rooms. This latter point is especially important when considering energy consumption. According to Fourier's Law, the rate of heat loss in a home is proportional to the difference between the internal and external temperatures (Lienhard 2011). To achieve a specific temperature (e.g. 20°C) in the daytime, when external temperatures are higher, lower rates of heat loss will occur than at night-time. As the thermostat triggers the boiler to come on based on internal temperature levels, greater heat loss at night will ensure internal temperatures drop at a faster rate, requiring the boiler to come on more often. Kempton (1986) was describing behaviour for householders who used the thermostat as their key control device. He found householders with a feedback model of the thermostat tended to have a usage pattern where they made infrequent, regular adjustments in line with routines of the household. Feedback model holders also tended to keep the thermostat set to a comfortable level at night, as they believed it would take more energy to heat up the cold bodies within the house if the temperature was allowed to drop too much. Due to the lower external temperatures, keeping a comfortable internal thermostat setting at night requires the boiler to work harder, that is to be on for longer periods, in order to compensate for greater level of heat loss. Kempton surmised that this limitation in the feedback model would result in systematically wasted energy. That is to say, sleeping householders who would be comfortable at a lower room temperature, maintain a higher set point when consumption rates are at their greatest.

    The chapter is structured to investigate three households known to hold a 'feed-back shared theory' of the home heating thermostat at the 'device' level. First, the households will be compared in terms of their energy consumption over a single week. The thermostat adjustment patterns for that week will then be examined to further understand the consumption. Next, the householders' self-reported typical adjustment 'strategy' with the home-heating controls will be used to help explain the patterns. Finally, the users' mental model description of home heating at a 'system' level will be examined to explain the strategy chosen. The implications of these findings for strategies to encourage reduced consumption will be dis-cussed throughout.

5.2   METHOD

This chapter uses data collected at the same time as that reported by a former study investigating variations in user mental models of the thermostat (Revell and Stanton 2014). The user group comprised overseas doctoral students at the University of Southampton, with young families, new to the United Kingdom. The participants were recruited by letter, email and approached door-to-door by the first author. Permission was sought from the faculty research ethics committee prior to con-tact and research governance was arranged (RGO Ref.8328). The participants were all from warm countries where centralized home-heating devices are uncommon. These user group characteristics ensured minimal prior experience of other home-heating devices, allowing behaviour patterns to be attributed to the mental model of the thermostat control installed, rather than habitual use from other devices (Revell and Stanton 2014). With mental models research, bias in data collection and analy-sis is a key consideration (Revell and Stanton 2012). The study took a number of steps to mitigate bias, including use of a carefully constructed questionnaire to gain data from the user with minimal leading to capture their mental models and typical behaviour. In addition, an analysis table was created for systematic categorization of householders' mental models of the thermostat as either 'valve' or 'feedback' shared theories (Kempton 1986). More details about the ways bias was mitigated and the procedures for data collection can be found in Chapter 3 and Revell and Stanton (2014). This section will describe the participants, settings and the methods employed for collecting the different data sources, and the choice of data representation.

5.2.1   PARTICIPANTS

The three participants in this case study were non-randomly selected from the Revell and Stanton (2014) study (Chapter 3). The basis for selection for this particular inves-tigation was for all participants to hold a 'feedback' shared theory of the thermostat (Kempton 1986). By controlling this variable, other causes for differences in behav-iour with the thermostat could be explored. Participant X was a student in his late thirties and originated from South Africa. He lived with his wife and two school-age children, and shared heating control with his wife. Participant Y was a postgraduate in his late thirties and originated from Taiwan. He lived with his student wife and young son and was in sole control of the central heating. Participant Z was a student in his early thirties and originated from Singapore. He lived with his wife and two young children, and took full control of heating for the family.

5.2.2 SETTING

The dwelling type, level of insulation, double glazing, location and central heating system were matched by using University of Southampton owned accommodation that had undergone identical refurbishment prior to the study, located on the same street in Southampton. This removed variations in thermodynamics of the dwelling or device design that otherwise may have influenced the behaviour and consumption levels observed. University accommodation was chosen for a number of reasons: (1) its convenient location when troubleshooting technical issues, (2) the opportunity to ensure the desired thermostat model was installed and (3) the ability to pre-install data collection equipment prior to residents arriving. The interviews were undertaken in a cafe in the university library where the participants could feel at ease.

5.2.3 DATA COLLECTION

Four key types of data were examined in this chapter. Data collected from the heating system including outputs of the system and inputs from the user. Outputs from the heating system related to energy consumption (boiler 'on' periods) and comfort (internal temperature). Inputs from the user focused on thermostat set point adjustments over time, as this was proposed by Kempton (1986) to be a significant behaviour variable that could result in systematically wasted energy. To help better understand and explain differences in consumption and behaviour, data were also collected directly from the user. This comprised users' strategies of home heat control and their mental model description of the functioning of the heating system.

5.2.4 FROM CENTRAL HEATING SYSTEM

The data collected from the heating system for this study were part of a larger study undertaken by the intelligent agents for home energy management group (IAHEM). The method for data collection was developed in conjunction with Hortsmann Ltd. Set point changes to the thermostat and room temperature were captured at 5 minute intervals. Data were transmitted from the thermostat via Z-wave radio waves to an Internet hub housed under the stairs in each household. The hub was connected to a 3G router which sent the data over a 3G mobile network to a secure server in Denmark (Seluxit). Data were extracted by colleagues in the IAHEM group from the secure Seluxit website (http://home.seluxit.com/). The set point and internal temperature data were calibrated by the IAHEM team with boiler 'on' times, resulting in data points per second. A single common week commencing 7 November 2011 for 7 days is examined in detail in this chapter. This allows illustration of how the different data sources selected can help explain energy-consuming behaviour. Differences in consumption between participants were represented with a single column graph showing the time the boiler was 'on' for each participant during that week (the mean thermostat set point was also shown for reference). Line graphs for each participant representing thermostat adjustments and internal temperature values (captured hourly) were constructed to compare actual behaviour. One graph per participant was produced combining these data with the corresponding period of time that the boiler was active.

5.2.5   FROM THE USER

In March 2012, following the period of automated data collection, the householder was interviewed to gain self-reported data of their behaviour with their home-heating controls. To capture each participant's strategy of home-heating control over a typical week, a simple template marking out the days of a single week (with an undefined y-axis) was annotated with the householder during a semi-structured questionnaire. For easy comparison, the structure of this output was designed to show changes in heating control adjustments over a week's period, to match the format produced by Kempton (1986) and the automated data collection outputs. In this chapter, these self-reported home-heating strategies have been redrawn for clarity.

To capture each participant's mental model description at the device and system levels of the heating system, questions and follow-on probes directed by a semi-structured interview template were used to build up a diagram with the participant. The diagram was created using post-it notes containing participant-initiated terminology relating to the heating system. Concepts were linked by drawing lines between the post-it notes. To gain insights into cause and effect, and rules of operation, participants were asked follow-on probes such as 'How does the boiler know when to come on/off' and 'What would happen if you turned the thermostat to its maximum setting?'. Participant responses to these probes were represented on the diagram using arrows and text. These mental model descriptions were then redrawn for clarity.

5.3   RESULTS AND DISCUSSION

Figure 5.1 shows the boiler 'on' periods for the three households with similar family setup, matched by location, house structure, insulation levels and home-heating technology, for a single week in November 2011. The mean thermostat set point for each home is also indicated, as this variable directly affects boiler 'on' periods.

The first thing to note about Figure 5.1 is the lack of correlation between boiler 'on' periods and the thermostat set point value. House X has the highest mean set point at 21.5°C, but not the highest boiler 'on' period. Houses Y and Z have the same mean set point of 20°C, but considerable differences in boiler on periods; the boiler is on in house Y for more than twice that of house Z. Guerra-Santin and Itard (2010) found the period of time the thermostat was set at its maximum value better correlated with energy consumption. Thermostat set point values over time would therefore better explain the differences seen in Figure 5.1. Kempton (1986) proposed that significant differences in consumption could be explained by differences in behaviour patterns of thermostat set points over time, where these differences were caused by users holding different mental models of the thermostat. Given the households mentioned all hold the 'feedback' shared theory of the thermostat, it was thought it possible that variations in behaviour patterns, arising from the same mental model, may account for these differences. To investigate this theory, the thermostat adjustment patterns were compared.

Figures 5.2 through 5.4 show the set point adjustment patterns, internal temperature and boiler 'on' periods.

Figures 5.2 through 5.4 reveal clearly that despite the participants for each house holding a feedback shared theory of the thermostat, only house X displays the infrequent, regular adjustments described by Kempton (1986). The readings from house X clearly show boiler 'on' periods matching increases in adjustment and boiler 'off' periods following decreases in adjustment, indicating the thermostat is a key part of their strategy for heating control. This is in contrast to the outputs from the houses Y and Z, which indicate no adjustment of the thermostat set point and very different patterns of boiler 'on' periods. House Y shows boiler 'on' periods only at night (approximately 9 pm–1 am), often continuously. The internal temperature reading shows that at no point was the desired 20°C set on the thermostat achieved, and the internal temperature sometimes dropped below 17°C during the day. House Z shows boiler 'on' periods 1–3 times a day between 6 am and 7 pm, and occasionally for a short time at 9 pm. However, the internal temperature was far warmer, ranging between 18.5°C and 21°C. House X is clearly from a different population than houses Y and Z with their approach to thermostat control, but manages to achieve comfort levels following the thermostat set point (Figure 5.2). Houses Y and Z, whilst sharing the same adjustment strategy and set point value for the thermostat, have realized very different outcomes. The former has achieved low levels of comfort with high levels of consumption (boiler on periods), whilst the latter has achieved good levels of comfort, with far lower levels of consumption. In Chapter 3, Revell and Stanton (2014) described how householders combine controls in various ways to manage their home heating. To understand where the control settings for other devices are responsible for these differences in boiler 'on' periods, diagrams of the participants' self-reported 'strategies' for home-heating control and the associated interview transcripts were referred to.

The typical strategies adopted by participant X are depicted in Figure 5.5. From their transcript, this participant stated their approach to heating the home as: 'to keep warm is one thing, and saving money was another concern......It's very optimised, our whole heating experience'. As no other devices are depicted, Figure 5.5 confirms the assumption following that derived from Figure 5.2 that thermostat adjustment makes up the sole strategy for controlling home heating. As such, we expect to find a 'feedback' behaviour pattern as described by Kempton (1986). The infrequent regular adjustment during the day to match the lifestyle is supported in Figure 5.5 and the following transcript:

> we would up the heat when we wake up. And then in the afternoon the kids go to school – they normally leave about eight o'clock – we will normally either turn it down... and then later on when they come home again we will up the temperature, because the kids normally get more cold.

Kempton (1986) expected a user with a feedback 'shared theory' not to turn down the temperature on the thermostat below comfort levels. Figure 5.5 shows that when the thermostat is adjusted last thing at night, it is returned to a comfortable 21°C. Further evidence that participant X is not aware of the impact on rate of heat loss of increased

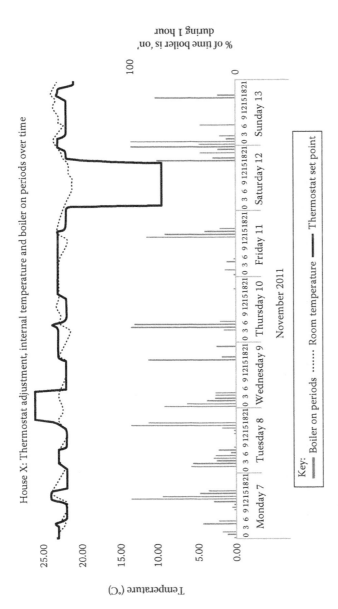

FIGURE 5.2   Remotely recorded thermostat set points, internal temperatures and boiler 'on' periods during a single week for house X.

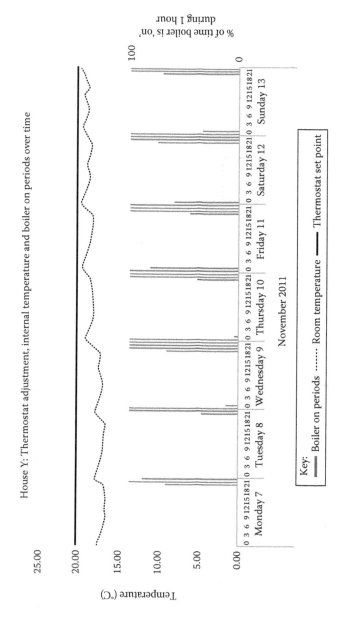

FIGURE 5.3   Remotely recorded thermostat set points, internal temperatures and boiler 'on' periods during a single week for house Y.

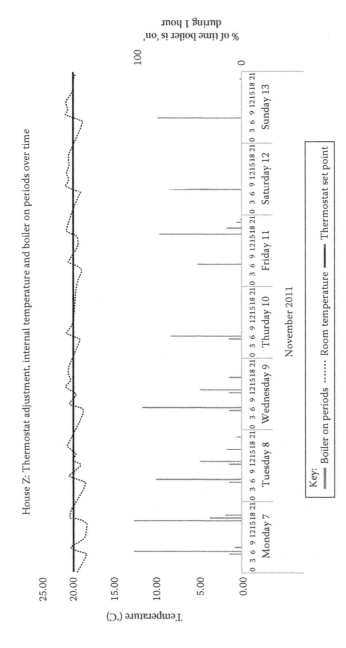

FIGURE 5.4   Remotely recorded thermostat set points, internal temperatures and boiler 'on' periods during a single week for house Z.

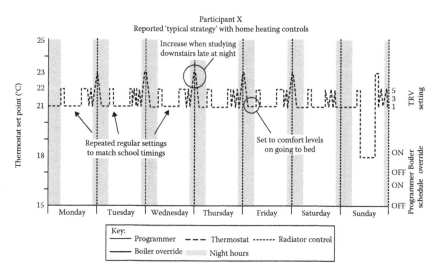

FIGURE 5.5 Devices used and typical adjustments made over a typical week reported by participant X.

temperatures at night is shown in Figure 5.5 with a 'spike' where the thermostat is turned up to 23°C for a short time. However, when referring to the transcripts, we can see that participant X does exhibit an awareness of infiltration of heat within the house, which encourages a strategy for heat control that wastes energy.

Analyst: Okay. So there might be situations, say, in the evening where you kind of put it up to 23 a little bit?

Participant X: Yes, for a short while because normally I sleep very late. I do about once or twice a week maybe I am at work until three AM maybe. So I would normally during this time put it up maybe for half an hour up to 23.

Analyst: And then back down to 21?

Participant X: Yeah. And then back to 21.

Analyst: To 21. So say you had it on for the evening around 22, you suddenly think 'Oh' – because you're in the cold room – you maybe put it up to 23, but then you turn it right down again?

Participant X: Yeah, because typically my habit would be to move from my desk..... and go to the front room and close the door, put it up on 23.

Analyst: So sort of heat up.

Participant X: It heats up and then I'll put it back on to 21

Participant X clearly has a good sense of how different parts of the house heat differently; elsewhere in his transcript he describes the front room as 'it's very warm, so you can't put it [thermostat] more than that [21°C] or you're feeling too hot'. By opting to work late at night in the coldest room (rather than the warmest) and raising the thermostat to maintain comfort levels in that room during the night, increasing

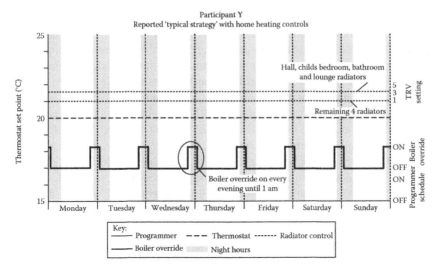

FIGURE 5.6 Devices used and typical adjustments made over a typical week reported by participant Y.

the rate of heat loss, he is getting 'poor value' for the amount of energy consumed on these occasions. Looking at the actual thermostat set points in Figure 5.2, we can also see that temperatures adjusted at night often remain on for longer than the 30 minutes stated, further extending boiler on times. When finally going to bed, as predicted by Kempton's (1986) feedback shared theory, this household does not 'set back' the thermostat but puts it at a comfortable level, where heat loss will be higher than during the day due to the greater temperature differential. Energy could be saved in this household with minimal impact on comfort purely by setting the thermostat to a lower than comfort level when sleeping.

Figure 5.6 shows that in addition to the constant thermostat set point, the thermostatic radiator valves (TRVs) and boiler override are both typically adjusted. From the interview transcript, participant Y describes his goal and strategy when managing the heating system:

> money is a very important driver for how to use the energy....we turn it [boiler override] on at about 9pm..and then we turn it off before we go to bed....But also we change the radiator control so typically right now we have eight radiators, but typically we only turn one bedroom, one living room, and one bathroom

User behaviour relating to night-time thermostat set points (Leung and Ge 2013) and TRV adjustment (Xu et al. 2009) contribute to increased consumption. The combination of heating only in the evening and night, when external temperatures are at their lowest, as well as using a limited number of radiators for heat transmission could explain the high consumption levels for participant Y. The lack of understanding of thermodynamics within a broader system supports Kempton's (1986) description of a user with a feedback shared theory of the thermostat. Lack of 'night set back' of the thermostat has instead been extended to the decision to put the heating on at night using the boiler override (which allows the thermostat to operate during this period). During the week discussed, the average internal temperature for household Y was approximately 18°C. The mean maximum and minimum external temperatures were 13°C and 9°C (www.wunderground.com), and using these as rough estimates for daytime and night-time temperatures, the temperature difference during the day would work out around 5°C, compared to 9°C at night. Choosing to override the heating at night instead of during the day could almost double the rate of heat loss, increasing the boiler on times to compensate for the greater loss. In addition to this, by reducing the outputs for heat transmission by half, unless all doors surrounding the hallway (which houses the transmitting temperature sensor) were closed, the boiler would again need to be on for proportionally longer periods to achieve the desired temperature. Had the heating been left on long enough to achieve the 20°C at night during this specific week, it may have needed a period of up to four times the duration necessary than that during the day, with all radiators transmitting. Money was also the driver for the thermostat set point choice as shown in the following quote:

> ...in the beginning we set it [thermostat] to 22, for maybe a couple of weeks...... and then some people [Newspaper advert] say that when you use set the temperature down one degree, some people say that you save 25 or 30 % of your costs.

However, given Figure 5.3 shows the heating was switched off before the boiler had achieved the thermostat set point, it is likely that little, if any, money was saved by making this adjustment. It is hoped that how advice to save energy would benefit from being context-specific has been illustrated. By actively trying to 'save money' through heat control, this household resulted in the highest boiler 'on' times of the case studies.

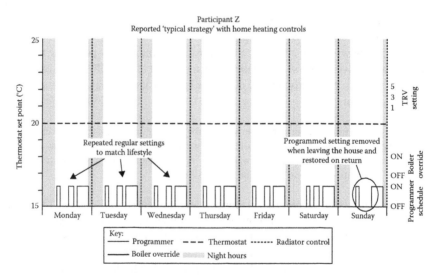

FIGURE 5.7 Devices used and typical adjustments made over a typical week reported by participant Z.

Figure 5.7 depicts the programmer as the main heating control used in conjunction with a static thermostat set point. Participant Z reported that his attitude to home-heating control was 'I want to keep warm'. They describe their usage of the thermostat as follows:

> I think it's about 6.50 am or 7.00 am it will switch on, and off at about 8.30 am.Then it will be on again about 12.30 pm, 1.30 pm, until 3.00 pm or something like that, and then it will come on again at around 5 pm to about half 10.....I will just change it according to what we are doing that day....[if going out for the day] I'd get rid of the middle one and depending on what time we expected to be home I would also adjust the second block.

This description is reminiscent of participant X's feedback strategy, with infrequent, regular changes based on the routines of lifestyle, but choosing to make the adjustment with a programmer rather than thermostat device. Participant Z does not describe a lifestyle that has frequent late nights, so the need to have the heating on when temperatures are lowest does not arise. In addition, as the programmer is used in place of the thermostat, the adjustment at 10.30 pm to turn 'off' the heating before bed circumvents this risk. By changing control devices and following lifestyle comfort needs, the largest risk to feedback model holders of the thermostat, excess heat loss at night is averted, without a need for participant Z to understand the effect of broader system thermodynamics on heating operation.

Participant Z did, however, demonstrate a sound understanding of the functioning of the thermostat, its dependency on its immediate environment, as shown in the following transcript excerpt:

Analyst: .... You mentioned here that the thermometer measures the hall temperature particularly. How does the radiator and other parts of the house affect the thermostat?

Participant Z: Not at all......Which is something that I didn't like, because of this I have to shut the corridors to control my thermostat....

Analyst: .... you can't necessarily control how much energy you use in the way that you want because you have to rely on the way it measures it?

Participant Z: Yes. It's the reason I switch on the radiator in the hall because I want to make sure that the boiler turns off when it reaches a certain temperature, not because I will stay in the corridor all the time.

Analyst: No (laughs). So the radiator in the hall is purely for operating the device not for your comfort at all, in your mind.

Participant Z: Yes.

Participant Z mentions shutting doors to the corridor and so is clearly aware of the effects on infiltration on the speed with which a radiator can heat up an air space. This understanding, whilst different from that suggested by Kempton (1986), who considered user comfort, is nevertheless significant in terms of energy consumption. By providing the appropriate conditions for the thermostat, the boiler can legitimately be told to turn 'off' more often, reducing consumption. The combination of: (1) choosing the programmer as the key control (so the user does not forget to turn it off), (2) scheduling heat use during the day (rather than at night) and (3) creating the appropriate environment to allow a central thermostat to operate effectively (closing corridor doors, keeping the hall radiator on) go some considerable way in explaining why participant Z has far lower levels of consumption than participants X and Y.

All three participants had been categorized as having a feedback shared theory of the thermostat, according to Kempton (1986). Two of the participants, X and Z, had clearly gone some way to 'adding on' a model of infiltration which benefited them in terms of comfort (e.g. participant X) or reduced consumption (e.g. participant Z). None of the participants described an understanding of the effect of differences in external temperature on boiler periods, which explains less appropriate use of the thermostat at night for participants X and Y, wasting energy. It is worth noting that these participants both mentioned saving money as a driver for heating control, yet both consumed considerably more than participant Z, whose aim was to keep the family warm. A willing attitude, therefore, is insufficient for reducing consumption, if how to effectively operate the heating system is misunderstood. A pre-requisite for promoting energy-saving behaviour is to provide energy-saving guidance for control devices *in the context of their environment.* Where the technology fails to communicate changes in consumption based on temperature differentials or infiltration, advice is needed to highlight these aspects for consumption to be reduced. Each participant also chose a different strategy in terms of controls used to operate their home heating. It was of interest how user mental models at a system level could explain the differences in the chosen strategy, so these participants' mental model descriptions of their home heating system were considered.

Figures 5.8 through 5.10 show the user mental model description of the home-heating system for participants X, Y and Z. There are clear variations in the complexity and elements present in the descriptions. In this section, we will briefly describe the function of the heating system relating to each model. This chapter will then focus on characteristics of each model that explain differences in the strategies adopted by each participant to control the heating system, and relate this to mental model theory to explain differences in the strategies adopted. In doing so, how the user mental model at a system level warrants consideration even when users hold the same device models of a key heating control will be illustrated.

Figure 5.8 shows the model description for participant X. How the heating system functioned according to this description is paraphrased as follows:

> The thermostat has a sensor that measures the room temperature in the hallway. It compares this room temperature and the temperature you have set on the thermostat. If there is a discrepancy, it will communicate to the box in the kitchen that the boiler has to go on. The light on the box in the kitchen will go green when the heating is on, and red when the heating is off. When green, water heats up in the boiler then travels to the radiators. You will then be able to feel the heat in the radiators. The thermostat will try to match the room temperature to the set temperature. When it has reached the set temperature it will send a signal telling the boiler to turn off.

The paraphrased description indicates a clear 'feedback' shared theory for the thermostat based on the description by Kempton (1986). Figure 5.8 describes the simplest model of the three participants, and shows the thermostat as the only control

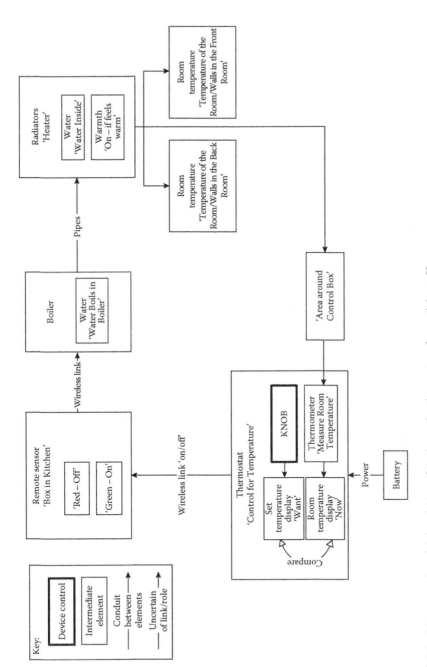

FIGURE 5.8   User mental model description of the home-heating system for participant X.

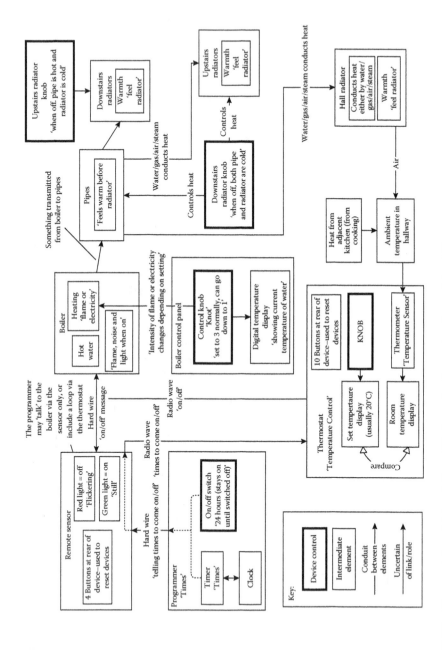

FIGURE 5.9   User mental model description of the home-heating system for participant Y.

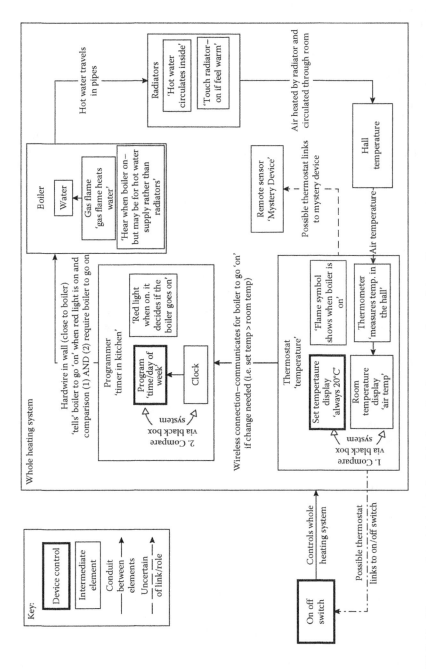

FIGURE 5.10   User mental model description of the home-heating system for participant Z.

device present. It is unsurprising, then, that this is the sole device to feature in participant X's strategy for heating control. This matches the findings in Chapter 3 and Revell and Stanton (2014). However, unlike the participant referred to in that study, participant X makes reference to a programmer device early in his interview transcript, but described it as 'too complicated' and 'too much trouble' to use. He did not refer to it when constructing his mental model description of the heating system. The importance of background knowledge in mental model theory, as a repository from which user mental models of devices can be inferred has been widely adopted in the literature (Johnson-Laird 1983, Moray 1990, Bainbridge 1992, Revell and Stanton 2012). The programmer clearly was present in the participant's 'knowledge base' (Bainbridge 1992), but did not make it to what Bainbridge describes as the 'working storage' (user mental model) referred to when operating the heating system. Different theories of mental models include different concepts (Revell and Stanton 2012). Bainbridge (1992) emphasized how meta-knowledge, which is knowledge and outcomes of the user's own behaviour, affects the strategies chosen when performing in complex systems. Participant X's difficulty at operating the programmer could have assigned 'meta-knowledge' to the internal concept of that device, preventing inclusion in the user mental model. This absence would then rule out selection of the programmer as part of a home-heating strategy, despite awareness of the existence of the device within the home. The problem of poor usability of home-heating programmers is clear from the literature (Combe et al. 2011, Peffer et al. 2011, 2013, Meier et al. 2011). Absence of programmers from user mental models of heating, when present in background knowledge, may be a widespread problem that gives an alternate explanation to the reluctance by users to revisit operating the programmer after initial frustrations. The strategy to make devices more intuitive is certainly positive. The provision of clear, immediate and user-specific information for operating programmers is also essential for energy savings to be realized (Darby 2001). However, householders who have a programmer absent from their mental model of home heating are unlikely to seek out changing their device to a different model to save energy. Overcoming negative meta-knowledge about the usability of the programmer is also necessary.

Figure 5.9, produced by participant Y, shows the most complex model description of the three participants. The functioning of the system agrees with that described by participant X, but is appended by a range of additional heating controls to the thermostat (Figure 5.9).

The boiler override is described as an 'on/off' switch and in Figure 5.6 it can be seen linking to the remote sensor. This then connects to the thermostat so that it knows during which times it can manage the boiler on/off times (in response to the hall temperature). The programmer is described as an automatic version of this boiler override, with the times for thermostat operation programmed into it. Figure 5.6 also shows a control knob in the boiler control panel that connects to the boiler and controls the intensity of the flame. This in turn affects the temperature of the water going through the pipes into the radiators. This understanding of the functioning of controls is essentially correct. However, participant Y misunderstands the functioning of the TRVs, the final control in Figure 5.9. According to participant Y, these controls allow hot water into the pipes or radiators when turned on, but block the water from entering the pipes or radiators when turned off. Participant Y appears to think of the TRVs as a type of manual on/off switch. The actual functioning of this device is akin to the hall thermostat. The TRV measures the temperature in the area surrounding the radiator. Each setting on the control is associated with a temperature range. When this temperature range is reached, the TRV slowly blocks the flow of hot water entering the radiator, so no more heat is emitted. When the temperature surrounding the radiator drops below the set temperature range, it slowly unblocks the flow of hot water entering the radiator, so heat can again be emitted. The literature describes how slow-responding systems are problematic. Operators of such systems usually achieve far below optimum performance (Crossman and Cooke 1974) and it is more difficult for users to form an appropriate mental model of function (Norman 1983). The strategy adopted by participant Y agrees with Brown and Cole (2009) that poor comprehension of heating controls leads to suboptimal levels of comfort.

With such a range of controls in their mental model, what is the reason for participant Y choosing the strategy in Figure 5.6? The data in Figures 5.1 and 5.6 refer to a specific week in November 2011. The strategy depicted refers to that time period, which was the start of the cold period, and the heating had not been used a great deal before this. It makes sense to use the boiler override, which requires no setup at this early stage of operation. As it got colder from late December, participant Y reports in his transcript that his strategy changed. The programmer was then set to control the heating to cover a 12 hour overnight period between 8 pm and 8 am Whilst we do not have boiler or thermostat data for participant Y beyond November, it is highly likely that the energy consumed will have significantly increased when the programmer was adopted as the main control, due to extended night-time operation, and the thermostat sensor positioned downstairs with a single radiator emitting heat. Participant Y described a strategy in Figure 5.6 of keeping only three radiators plus the hall radiator turned on, and the others turned off. The impetus for this strategy was to save energy in rooms that are not occupied for long periods. However, this strategy is only effective if the rooms where the radiators are turned off have their doors permanently closed. Otherwise, filtration of warm air from the heated rooms to the colder rooms will lower temperatures in the heated rooms, so the radiators will need to emit more heat. Whilst Figure 5.9 does indicate that the hot kitchen heats the ambient temperature of the hall, he does not refer in his transcript to infiltration to other rooms lowering the temperature in the hall, nor emphasize shutting doors as part of his strategy (as described by participant Z). Participant Y described his strategy for adjusting radiator controls when using the programmer as follows: During the day, all radiators (apart from the hall) were turned off. During the evening, the downstairs radiators were kept on, and when going to bed, the downstairs radiators were turned off and the upstairs radiators were turned on. This level of control assumes a faster responding control than the TRVs installed in this home. As such, and as was seen in Figure 5.3, the majority of the time, the internal temperature would be below comfort levels, regardless of the thermostat set point.

Moray (1990) proposed an alternate view to Bainbridge (1992) regarding the mechanisms behind mental models of complex systems, which may explain differences in home-heating strategies. He related the organization of information in mental models in terms of lattice theory, and proposed that operators selected strategies associated with Aristotle's four causal hypotheses. A lattice containing a hierarchy of cause-specific information would be derived from the background knowledge of the system. Depending on the strategy chosen (formal, material, efficient or final), the operator would limit reference to information in the associated lattice (until a change of strategy was adopted). It is proposed that different attitudes to home-heating use could be interpreted in terms of different causal strategies, resulting in focus on the control devices present in the associated lattice. For example, participant Y's determination to save money by saving energy would trigger reference to an *efficient* cause (by considering ways the heating system can confine and limit heat from being emitted), whereas participant Z's focus on 'keeping warm' would be classed under a *final* cause (as he is only considering how the heating system can provide comfort). Participant Y's determination to control heat flow may have motivated an interest in home heating that resulted in a greater number of devices being present in his background knowledge from which to infer the mental model description in Figure 5.9.

Participant Y also describes adjusting the boiler water control as atypical, such as when going on holiday during winter, as part of a strategy to prevent the pipes from freezing. By having a more complete mental model of the heating system, the options for different strategies to suit different circumstances are increased. However, it also presents a risk of inappropriate strategies being adopted, particularly if the device model for a single control is inaccurate and how the heating system operates within broader systems is not understood.

Figure 5.10 shows the mental model description for participant Z. The mental model description broadly agrees with that provided by participant Y, with the addition of a programmer device and on/off switch to the whole system. The remote sensor is represented in this description as a 'mystery device', as participant Z was unsure of its function. However, as the remote sensor was not a control device, merely an intermediary for communicating to the boiler, the omission of this element does not impact the strategies chosen.

Unlike the description provided by participant Y (see Figure 5.9), control devices such as the boiler temperature control, boiler override and TRVs are absent from the description in Figure 5.10. The lack of awareness of these devices explains why they are absent from the typical strategies shown in Figure 5.7. From the transcript, participant Z is a bit unsure of the actual role of the power switch, but imagines, correctly, that this is an electrical isolation switch required for the heating system to operate. He does not consider it as a heating control in the same way as the thermostat and programmer, so this does not form part of his typical strategy for controlling heating. Using Moray's (1990) lattice theory, the power switch would form part of a *formal* cause strategy (the system can operate when it is in the 'on' position), rather than a *final* cause relating to providing comfort in the home.

Moray's (1990) perspective on how strategies are chosen provides some interesting implications for the design of technology or provision of energy-conserving advice. If, as in this case study, householders with an 'efficient' cause strategy are more likely to have complex mental models for home heating (featuring a large number of controls), then technology that helps promote the way different control devices work together could help realize energy conservation goals. Simplified 'single' device energy-saving advice is clearly inappropriate for this audience. A simplified message may be most appropriate for an audience with a 'final' cause strategy, who may be more predictable in the way they follow advice.

5.4   SUMMARY AND CONCLUSIONS

In summary, we have seen how mental model theory, and differences in the topology of the mental model description of the heating system, can play a useful role in explaining differences in strategies chosen, despite users holding the same 'feedback' shared theory of the thermostat. In this case study, the larger number of control devices evident in the description led to strategies involving more control devices. Inappropriate mental models at the device level for alternate heating controls (such as the TRVs with participant Y) were shown to enable inappropriate inclusion in a home-heating strategy – in this case, contributing to wasted energy and inadequate comfort levels. Kempton (1986) described limitations to users with a 'feedback' shared theory to include an understanding of the effect of infiltration on comfort and external thermodynamics on rate of heat loss. The interview transcripts brought to light that the effect of infiltration on appropriate functioning of a central thermostat (rather than comfort alone) is highly pertinent. Where this understanding is missing, and other devices that affect heat output make up part of the householders' strategy (e.g. TRVs), there is a further risk of wasted energy. Whilst two of the participants mentioned infiltration in this sense in the interview transcripts, there was no evidence in any of the transcripts of an understanding of external temperature on rate of heat loss. That two of the participants had opted to use heating late at night when external temperatures are lowest, corroborates Kempton's (1986) view that energy can be systematically wasted with a broadly accurate device model of the thermostat. Where strategies combine thermostat use with the programmer, energy use may be reduced (as with participant Z, where the heating is set to turn off before bedtime), or increased (as with participant Y, who described extending night-time heating after incorporating the programmer into his strategy). Similar variations have been found by Peffer et al. (2011) and Shipworth et al. (2010). It is clear that householders' mental models of home heating need to include the impact of external temperatures for significant 'systematic' energy savings to be made. Both, technology and energy-saving advice could play a big role in this endeavour, by communicating appropriate behaviour with controls in the context of the environmental setting.

Understanding device models is useful, especially where the limitations affect energy consumption. However, patterns of behaviour may not be limited to the device under consideration. Understanding of user mental models at the system level allows appreciation of why some combination of devices are used, and where these can lead to a positive/negative effect on energy consumption. Mental model theory can help to understand why specific devices within mental models at a system level are chosen to form strategies of home-heating control. Efforts to understand or mitigate for inappropriate use by looking at outputs from single devices (e.g. thermostat, boiler and programmer) carry a risk, particularly for householders who have complex mental models. The implication for technology and energy-saving advice is to promote in the householders a 'systems view' of home-heating control (Revell and Stanton 2014).

Efforts to advise householders on ways to reduce consumption have shown to be ineffective at ensuring changes in behaviour (Darby 2001), even when delivered via home visits (Revell 2014), or even when advice has been actively sought by householders (Mahapatraa et al. 2011). Darby (2001) identified a considerable amount of wastage in standardized advice packages, with only some aspects relevant for individual households and confusing information resulting in none of the advice being acted on. A growing body of research points to the need for clear, tailored advice that enables householders to act on intentions to reduce consumption (e.g. Darby 2001, Fischer 2008, Kuo-Ming et al. 2012) and approaches have been developed to use technology as a means for providing different forms of custom advice (e.g. Shah et al. 2010, Shigeyoshi et al. 2011). This chapter has illustrated how householders with a Kempton (1986) feedback shared theory of the thermostat (that overlooks the influence of thermodynamics and heat flow in a broader system) can unintentionally waste more energy whilst endeavouring to conserve it. The findings suggest that tailored advice is required, whether through government campaigns or technological innovations. This advice should consider variability, not just in demographics and attitudes, but in the thought processes that translate intentions into actions. In doing so, promised energy savings may finally become a reality.

Chapter 6 takes this sentiment one step further and looks at how variability in user mental models of home heating could be used to form a design specification for home-heating interaction that promotes greater consistency in mental models with a view to encourage appropriate behaviour with home-heating controls.

6 Mind the Gap

A Case Study of the Gulf of Evaluation and Execution of Home-Heating Systems

6.1 INTRODUCTION

This chapter provides a methodology to facilitate investigation of hypotheses 3 and 4 described in the introduction of this book. Using data collected by the QuACk method described in Chapter 4, it uses Norman's 'gulf of evaluation and execution' to determine design elements that should be emphasized to promote a mental model of home heating that would allow appropriate behaviour with controls. This provides a starting point for investigating how device design influences users' mental models (hypothesis 3), as well as the overall aim, how design can be used to influence mental models of home heating, to encourage patterns of device use that influence the amount of energy consumed over time (hypothesis 4). The findings from this chapter feed into design decisions made in Chapter 7 to create a novel home-heating interface simulation to test if user models are altered as intended in Chapter 8.

Reducing domestic energy consumption is one way the United Kingdom aims to fulfil legislation requiring greenhouse gas emissions to be cut by 80% by 2050 (Climate Change Act 2008). Home heating contributes 58% of domestic energy use in the United Kingdom (Department of Energy and Climate Change 2011). Considerable differences in the amount of energy used in homes result from the occupant's behavioural differences (Lutzenhiser and Bender 2008). Understanding the cause of behavioural differences in the way home-heating systems are used, provides an opportunity to develop approaches to reduce domestic energy consumption. The purpose of this chapter is to provide insights that could inform the design of a home-heating digital control interface to encourage appropriate heating control. 'Appropriate heating control' is considered, in this book, to be pragmatic operation of heating devices so householders fulfil their heating goals with minimal wasted energy. The approach taken in this chapter is to apply Norman's (1986) theory of the gulf of evaluation and execution to data collected on mental models and typical behaviour relating to a domestic gas central heating system.

Home heating is considered by Sauer et al. (2009) to be the most complex system in the domestic setting. The domestic setting varies considerably in its occupants, their energy and heating needs (Stern and Aronson 1984, Lutzenhiser 1993), and therefore heating goals. This diversity encompasses households that vary in structure, level of insulation and external temperatures, all of which affects the thermodynamics of the dwelling (Kempton 1987). There are also considerable variations in the heating system equipment within peoples' homes, particularly due to their modular nature. This allows wide-ranging combinations of thermostats, programmers, thermostatic radiator valves (TRVs), boilers, radiators and heat sensors. Different configurations can result in differences in the type and range of heating solutions available. Due to the high cost of equipment and installation, householders typically inherit a heating system configuration when they buy or rent a house. This means there is no guarantee that householders' particular goals can be satisfied by the existing system. These multiple variations complicate approaches to reduce consumption that intend to put consumers 'in control' of the heating system. With so many ways of controlling the heating system, it is difficult for householders to understand how to achieve their heating goals without wasting energy. With variations between households in the system setup, it is difficult for government agencies to offer clear and effective 'one-size-fits-all' advice to householders to realize this end. Government campaigns in the United Kingdom (e.g. www.energysavingtrust.org.uk) tend to focus on energy reduction. Guidance and assistance to reduce consumption that recognize variations in householders' heating goals, their capabilities and the capabilities of their heating system, will be more eagerly adopted and maintained long-term.

Key issues that prevent householders operating heating systems appropriately include the cognitive and physical usability of the system. Kempton (1986, 1987) proposed that variations in the way people operate their home-heating thermostat, resulted from their differing 'mental models' of the way the device functioned. Kempton (1986) found evidence that behaviour patterns associated with some mental models were more energy-efficient than others, as they encouraged 'night set back'. In Chapter 4, Revell and Stanton (2014) found that faulty or incomplete mental models explained non-optimal operation of home-heating devices, where energy was either being wasted, or heating goals failed to be achieved. Considerable energy savings could be made if heating systems were effectively programmed (Combe et al. 2011, cited Gupta et al. 2009), yet in a study by Combe et al. (2011), 66% of participants were unable to complete the set programming task. Peffer et al. (2011) discuss how more energy can be wasted by incorrectly programmed heating controls than if their manual alternatives had been used. Users are not 'in control' of their heating system in the way manufacturers intend, if they misunderstand how home-heating systems contribute to their goals, and find it difficult to operate the heating controls.

One way technology has been adopted to assist householders with the management of their domestic energy use is the development of third-party interfaces. These engage users by providing feedback on their consumption (sometimes in context to other users), and allow varying degrees of digital control over their physical systems. Originally focusing on electricity-consuming devices (e.g. Alertme), more recent offerings now target remote home-heating control (e.g. British Gas Hive Active Heating, Tado, Honeywell Evohome), or use intelligent automation of home heating (e.g. Tado, NEST). Each of these applications takes different approaches to the way information is presented and the type of control possible, and was designed to achieve different objectives. Whether these approaches support or hinder appropriate consumption would depend on the variations in the heating system, house structure and householders' goals.

To gain insights that could help specify the needs of an interface to support appropriate consumption from the perspective of overcoming cognitive and physical usability issues, this chapter looks to Norman's (1986) theory of the gulf of evaluation and execution. Norman (1986) introduced the idea of the 'gulf of evaluation and execution' to explain why computer users did not operate systems in the way system designers intended. Norman emphasized that this problem specification applied equally to physical systems, directing consideration of home heating as a suitable domain for application. Norman approximates seven stages of user activity to describe how the user bridges the gulf. These stages take into account user goals, their perceptions, intentions and actions. Norman (1986) emphasized the need for user mental models to be compatible with the design model of the system to effectively bridge the gulf of evaluation and execution. Norman (1986) supports the view of Buxton (1986), who highlights how the usability of the input controls can play a significant part in influencing the choice of control device operated. This has particular relevance for home-heating control devices, given the number of different control devices that can make up the heating system, and the range and variations in design of each device.

6.1.1 Norman's (1986) Gulf of Evaluation and Execution

Norman's gulf of evaluation and execution represents the distance between a user's psychological goals (e.g. I want to be warm whilst watching TV with my spouse in the living room) and physical actions necessary, with a specific system, to achieve those goals (e.g. press the boiler override button on my programmer). According to Norman (1986), the user bridges these gulfs by going through a number of stages (Figure 6.1). A user bridges the gulf of execution by (1) forming their intention to use the system to achieve their goal, (2) specifying the action sequence that will achieve their goal and (3) executing the necessary actions with the input devices. They also need to bridge the gulf of evaluation by (1) perceiving the state of the system, (2) interpreting the state of the system so it can be compared to their goal and (3) comparing the system state to their goal. Combined with goal specification, these stages make up Norman's (1986) 7-stage model for user activity, illustrated in Figure 6.1 with home heating as the context. This model shares similarities with Rasmussen's (1983) 'decision ladder' concept. If people are using home heating in a non-optimal way, it suggests that it is difficult to use the heating system in an optimal way. The 7-stage model for action may bring insights into why people fail to appropriately manage their home-heating systems. These insights could point to design requirements that may help reduce wasted energy.

Stages 2–4 bridge the 'gulf of execution' and stages 5–7 bridge the 'gulf of evaluation'. Norman (1986) suggests that users will be more successful at achieving their goals when interacting with systems, if efforts are made to facilitate users at each of the seven stages of activity. This can be achieved either by bringing the user closer to the system (through experience, or training) or by bringing the system closer to the user (Norman 1986). In the home-heating context, the designer does not control the level of experience of the user, nor can they demand they undergo home-heating training. Design-based approaches to 'bring the system closer to the

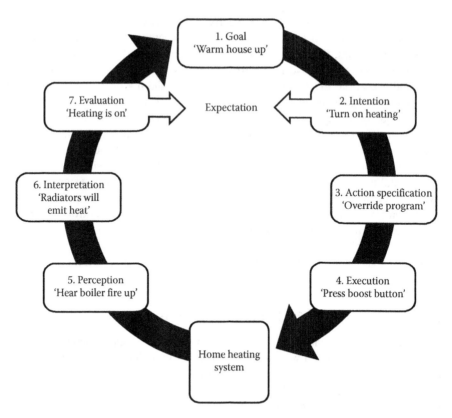

FIGURE 6.1   Norman's (1986) seven stages of user activity applied to home-heating context.

user' to effectively bridge these gulfs may result in more appropriate use of home-heating systems by householders. Evidence of this approach has been offered by Staffon and Lindsay (1989) in the design of power plant processes, where provision of an interface containing a thermodynamic model of plant performance ensured operators had a compatible 'mental model' of the system states, before executing the controls. Similarly, Connell (1998) concluded that the gulf of execution and evaluation had advantages over analysis methods that considered taxonomies of error (e.g. Meister 1977) or skills and rules (e.g. Reason 1990). When performing error analysis on ticket-vending machines, Connell (1998) found that viewing erroneous button presses within the framework of the gulf of evaluation and execution more simply explained causes of error and pointed to practical design solutions. The gulf of evaluation and execution has also been used to inform improvements to direct manipulation interaction (Hutchins et al. 1985, Mohageg 1991, Kieras et al. 2001), understand difficulties in using programming languages (Ko et al. 2004, Edwards 2005) and to facilitate human–robot interaction (Scholtz 2002). Norman (1986) believes many systems can be characterized by how well they support the seven stages of action. Cuomo and Bowen (1994) applied three different evaluative methods to determine how well they addressed issues at each of the seven stages of user activity. Cuomo and Bowen (1994) found these methods better addressed the stages that made up the gulf of execution (intention, action specification, execution) than those associated with the gulf of evaluation (perception, interpretation and evaluation).

Whilst systems have been evaluated from the perspective of the gulf of evaluation and execution to identify or explain execution issues, this has generally been done at the system interface or behaviour level. This chapter takes a different approach, by focusing on the users' mental models.

6.1.2 Conceptual and Mental Models of Home-Heating Systems

Mental models are internal constructs that are considered important in predicting, understanding and explaining human behaviour (Craik 1943, Johnson-Laird 1983, Wickens 1984, Kempton 1986). The notion has proved attractive when considering interface design (Carroll and Olson 1987, Williges 1987, Norman 2002, Jenkins et al. 2010), to promote usability (Norman 2002, Mack and Sharples 2009, Jenkins et al. 2011) or enhance performance (Stanton and Young 2005, Stanton and Baber 2008, Grote et al. 2010, Bourbousson et al. 2011). The definition of mental models has been the topic of much debate over the last 20 years (Wilson and Rutherford 1989, Bainbridge 1992, Richardson and Ball 2009, Revell and Stanton 2012).

In this chapter, Norman's (1983) definition of mental models will be used. He distinguishes between user mental models (UMMs) and conceptual (or 'design') models. The UMM is defined as 'the actual mental model a user might have', gauged by observations or experimentation with the user (Norman 1983). The conceptual (or 'design') model is defined as the model which is invented by designers to provide an accurate, consistent and complete representation of the system. For a fuller discussion of the distinction between these and other definitions of mental models, see Revell and Stanton (2012).

For effective bridging of the gulfs, Norman (1986) demands the user mental model is compatible with the design model of the underlying system. Norman (1986) proposes that the designer can promote compatible user mental models through the choices they make when constructing the system image, which in turn influences user mental models (see Figure 6.2).

Norman (1986) considered the conceptual model (or design model) as the 'scaffolding' for the bridges that enable users to cross the gulfs of evaluation and execution. If this underlying structure is unsound, or misplaced, there could be an impact on the seven stages of activity, ultimately affecting the way users respond to the system design at higher levels (see Figure 6.3). As a result, well-intentioned design efforts may fail to support users, or worse, confuse or mislead them. It is therefore of interest to identify where problems with the mental model 'scaffolding' of the home-heating system could cause difficulties that prevent appropriate operation.

This chapter refers to data collected from six householders from dwellings with identical structure and central heating systems, using the Quick Association Check (QuACk). This method is described in detail in Chapters 3 and 4 and was designed to explore association between mental models and behaviour. The method provides outputs in the form of mental model descriptions (similar to concept maps),

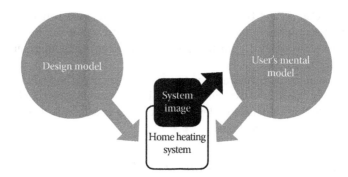

FIGURE 6.2   According to Norman (1986), the system image contributes to the user's mental model, influencing their interaction with the heating system. Appropriate operation is supported if the user's mental model is compatible with the design model of the heating system.

graph-based outputs of typical usage and interview transcripts. QuACk was also used to collect data from an expert in home-heating systems from the heating control manufacturing company, who designed the devices installed in the households. These data were used to represent a 'design model' and 'recommended execution' of the home-heating system, providing comparable data with which to compare the householders' data.

This chapter is structured to first illustrate the design model and compatible user mental model that would facilitate the recommended execution. How the seven stages of activity are bridged using the design model is then described. Next, the home-heating 'system image' will be considered to understand the ways it supports or undermines householders' development of a compatible mental model that enables appropriate execution of home-heating controls. Following this, the results of user mental models of householders in our case study, and their behaviour patterns, will be compared to that proposed by the expert. The seven stages of activity will be used as a basis for discussion to help understand why differences arise. The insights gained will be used to detail how key features of the user interface impacts users' mental models. The effects of this impact on home-heating and energy-saving goals at different stages of activities are tabulated in Table 6.5.

6.2   THE DESIGN MODEL

The design model is the conceptual model of the system (Norman 1986), so will be determined by the designer and developers of the system. As discussed in the introduction to this chapter, the home-heating system is a composite of a number of different components (boiler, programmer, thermostat, etc.), often made by different manufacturers. All of these different components have their own 'design model' at the component level, as well as a composite design model at the system level, based on how the components are combined within a household. Whilst design models of specific components could be inferred from user instruction and installation manuals, these tended to discuss device function in isolation, or at most, with very general

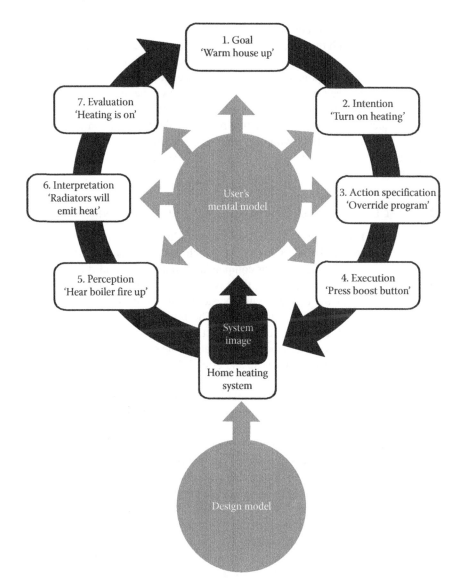

FIGURE 6.3 The compatibility of the user's mental model to the design model at each of the seven stages of activity characterize the 'structural integrity' of the bridges that span the gulf of evaluation and execution.

reference to connecting devices. This presented a problem when seeking a design model for the central heating system as a whole, which could be compared with user mental models of a specific system.

To overcome this difficulty, an expert that could give an overview of a specified central heating setup was interviewed. The expert approached had 40 years' experience with home-heating systems and worked for the manufacturing company that produced the key control devices present in the case study households. To enable direct comparison with the householders' data, the expert was interviewed using the QuACk method (Chapter 4). For reference, the expert was provided the layout and specification of the components present in the case study households, so he could keep this in mind when describing his mental model of the system. To provide an example of 'appropriate' behaviour with this heating system, the expert was asked to respond to questions, as if he was making recommendations to a member of a family with school-age children, with one adult at home during the day, and the other working outside the home. This reflected the family structure of householders in the case study.

6.2.1 THE DESIGN MODEL EXPRESSED AS AN EXPERT 'USER MENTAL MODEL'

Figure 6.4 depicts the mental model description produced using the QuACk method (Revell and Stanton 2014 and Chapter 4), which has been redrawn for clarity. It shows at a system level the different components, connections and rules associated with the central heating system from the case study. Each box in Figure 6.4 represents a component of the central heating system with the benefit of the system to affect 'room temperature' (in dotted lines). The boxes outlined in bold represent home-heating controls that a user could interact with. The boxes with a light outline represent ways in which devices indicate the state of the system. The arrows between boxes represent the direction of 'cause and effect'. Rules and functions of elements and links are annotated with text.

Figure 6.4 illustrates an interrelated relationship between heating controls and the system benefit. This design model is far from straightforward, but Norman (1986) does not consider it necessary for a householder to understand the full complexity of the design model of a device, in order to use it effectively. Norman (1986) does require that the user mental model is 'compatible' with the design model, however. Following discussions of appropriate operation with the heating system, the home-heating expert was asked to identify which parts of the design model shown in Figure 6.4 he thought were necessary. That is to say, what 'compatible' mental model would allow a householder to operate the system appropriately. These are identified in Figure 6.4 with bold italics and tabulated with explanations in Table 6.1 (rephrased where appropriate). It should be emphasized that as the opinion of the home-heating expert, these components provide a starting point, rather than prescriptive guidance on how to build a compatible user mental model of a home-heating system.

To provide a visual example of what a compatible user mental model to the 'design model' would look like, the essential components of the design model in Figure 6.4 have been redrawn in a simplified form in Figure 6.5.

Figure 6.5 depicts (from the left) the master switch that supplies electricity to the heating system, linked to the boiler. It shows the boiler with two controls attached to it, the programmer (with schedule/boost/advance features) and the thermostat control. The programmer schedule calls for heat to the boiler at specific times. It has a boost button that adds an ad hoc single hour to the schedule from the time pressed, and an advance button, that will set the next part of the schedule to start immediately. The boiler has a conditional switch that requires the programmer and the thermostat

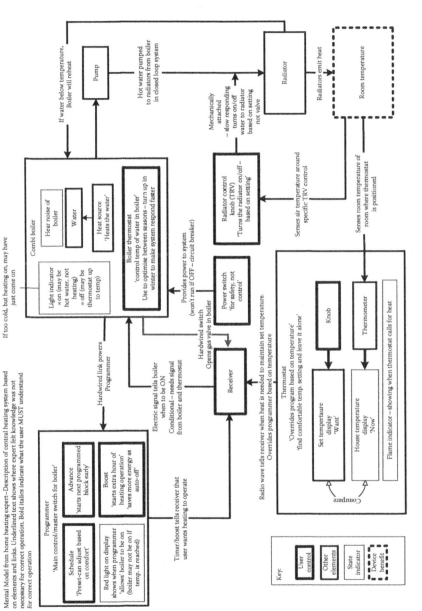

FIGURE 6.4 The home-heating system 'design model' represented as an expert user mental model description.

TABLE 6.1
Expert Considered 'Essential' Components of Compatible User Mental Model for Appropriate Operation of Heating System

Essential Components	What User Needs to Understand for Appropriate Operation
Programmer schedule	Master control for the heating system
Programmer boost	Override for schedule that automatically switches off after 1 hour
Thermostat	Only calls boiler for heat when set point temp > temp of room
Room temp	Only room temp where thermostat is situated is used to control boiler
Boiler	Heats and pumps hot water around system
Conditional rule in Boiler	To operate, boiler requires both programmer AND thermostat to call for heat
Pipes	Contains hot water in closed loop system (helps user understand operation)
Radiators	Emit heat
Thermostat radiator Valve (TRV)	Not work like a valve – not for frequent adjustments – slow response control, used for limiting heat long-term, turns radiator on when set point > sensed temp, and off when set point < sensed temp
Master on/off switch	Not for heating control, but safety (electrical isolation) when maintaining system. Provides power to the system

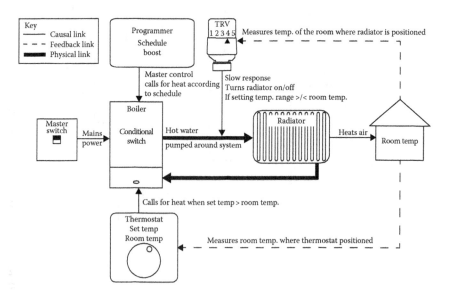

FIGURE 6.5   The elements of the design model that should be evident in compatible user mental models of a home-heating system.

to both be calling for heat before it will pump hot water around the pipes to the radiators. The radiators emit heat to the air in the room, resulting in a change in room temperature. The thermostat takes an estimate of the house temperature from the room where it is positioned. It compares this room temperature to its set temperature value, calling for heat from the boiler only if this is higher. Before reaching the radiators, the TRV can control if the hot water will enter the radiator. This is a crude slow response control and works like the thermostat. If the room temperature measured by the TRV is higher than the temperature range associated with its set point, then it will allow hot water to flow into the radiator. If lower, the TRV will slowly block access, so hot water must bypass that particular radiator, and that room will no longer continue to receive heat from the radiator.

6.2.2 WHAT DOES 'APPROPRIATE' HOME-HEATING CONTROL LOOK LIKE?

What an expert in home heating considers a 'compatible user mental model' to look like has been shown. Based on the behaviour output of QuACk, Figure 6.6 depicts the expert-recommended use of heating controls over a typical week in winter. This reflects the desired state of the system that would result from stages 3 and 4 of Norman's (1986) seven stages of activity (how householders specify and execute their heating goals with their home-heating system). The recommendation will represent a 'standard' to which householders' reported behaviour will be compared.

6.2.3 STAGES OF 'APPROPRIATE' ACTIVITY WITH A HOME-HEATING SYSTEM

Stage 1 of Norman's (1986) seven stages of action represents goals. The heating goals assumed by the expert for a family with young children were to (1) enable the house to be comfortable during routine times (e.g. in the morning on waking and getting ready to leave the house, and at the end of the day when returning and relaxing before bed), (2) to have ad hoc comfort on demand when the house is occupied (e.g. by parent during the week, or whole family at weekend) and (3) to avoid wasting money on heating the home when it is not needed (e.g. when the house is unoccupied, or when occupants have high activity levels, doing exercise or housework). Norman's (1986) second stage of action, intentions, is closely related to goals and represents the decision to act, to achieve the goal. The intentions described by the expert were to use specific control devices of the central heating system to appropriately achieve these goals. For goal 1, the expert's intention was to set the long-term heating controls and for goal 2, the intention was to override these pre-set heating controls. The intention to avoid heating the home when not needed (goal 3) was to utilize residual heat, to avoid heating long-term unused rooms and to check if comfort levels are the result of the room temperature, or activity levels, before deciding to override the system. How Norman's (1986) seven stages of activity relate to these three home-heating goals is summarized in Figure 6.7.

Stages 3 and 4 involve mapping the user goals and intentions onto a desired home-heating 'system state'. In this context, it represented the required adjustments of the programmer, thermostat, boost and TRV to fulfil the goals, and the actions necessary for execution. Figure 6.6 depicts stage 3, the intended 'state' of

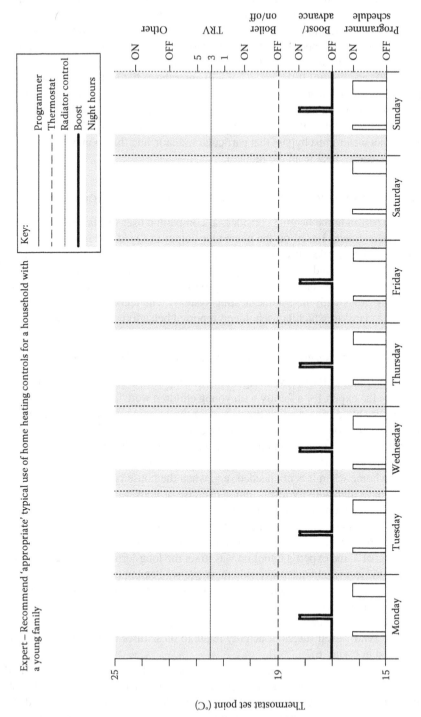

FIGURE 6.6 Recommended state of the home-heating system – stage 3 of Norman's seven stages of activity.

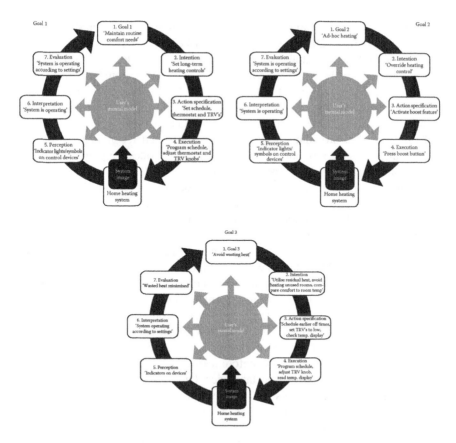

FIGURE 6.7 Norman's (1986) seven stages of activity broken down by typical home-heating goals.

the home-heating system controls following this execution. To fulfil goal 1, the expert recommended 'action specification' was for the programmer to be used in conjunction with the thermostat and TRVs. At the bottom section of Figure 6.6, the programmer timeslots are set from 1 hour before the household awakes during weekdays, to enable a comfortable temperature on waking. In the evening, the programmer time slot is set from 30 minutes to 1 hour before the occupants return from work or school, allowing time to reach a comfortable temperature when the house will be occupied. The expert specification was to program a time slot for 4–5 hours (depending on when the occupants go to bed). Figure 6.6 shows this pattern repeating daily during the week, and with time slots shifted slightly later at the weekend as an example of 'appropriate' heating operation with the programmer to fulfil routine comfort needs (goal 1). It is important to make clear that the programmer time periods do not represent 'boiler on' time periods. Heat energy is only emitted during these periods when the thermostat also calls for heat. As Figures 6.4 and 6.5 show, the programmer works in conjunction with other devices. According to the expert, industry recommendations state the thermostat set point should remain static at around 18°C–20°C (see Figure 6.6, dashed line). Due to variations in the structure of the houses, initial experiment is necessary to determine a set point that achieves a comfortable temperature in the most-used rooms. Similarly, the TRV set points should largely be kept at static settings and should not be treated as a heating control, but a heat limiter. The expert action specification was for used rooms to adopt a moderate, static setting (see Figure 6.6, light grey line).

To fulfil goal 2, providing ad hoc comfort, the expert recommended using the 'boost' button (Figure 6.6, bold line). This control overrides the program schedule, inserting an instant extra hour where heat can be emitted (subject to the thermostat). Operating the boost does not interfere with any other long-term device settings, so the user does not have to reinstate original settings after use.

To fulfil goal 3, the expert specification relates to the other goals. When setting the programmer for goal 1, to fulfil routine comfort needs in the morning, the expert recommends the time slot end 30 minutes after waking (see Figure 6.6), as the comfort levels will be retained without further heating for the next hour (assuming reasonable insulation), keeping the home comfortable until they leave for work and school. This avoids wasting energy by retaining heat in an unoccupied house. When making long-term TRV settings, the expert recommended adjusting TRVs in long-term unused rooms to a low static set point. When considering overriding the heating to fulfil goal 2, the expert strongly recommended that comfort levels should be compared to the air temperature before taking action, by referring to the 'room temperature' display on the thermostat device. Often, low activity levels, hunger or tiredness may make individuals feel cold, when the heating system has achieved the intended air temperature. The expert considered that energy is wasted if the air temperature is raised to compensate for these other variables, which would be more appropriately addressed directly (e.g. by increasing activity levels, eating, sleeping). The recommendation for using the boost (as opposed to other override options, such as on-off button, turning up the thermostat) also is linked to goal 3. The expert emphasized that the feature of automatically switching off after 1 hour prevents energy being wasted by being on for longer than needed through the risk of users forgetting to revert to the original settings (e.g. thermostat being left at a high setting, or on-off switch left to on).

The expert also emphasized that a benefit of opting to use the boost, is the ease in which the user can activate the setting (by pressing a button). This relates to stage 4 of Norman's (1986) stages of activity – how a user executes action specifications. To execute adjustments to the program schedule (for goals 1 and 3) required the householders in the case study to make various mode changes and adjustments on the programmer device. This was an involved process that would need the user to refer to instructions. To adjust the set temperature value on the thermostat (for goal 1), householders merely had to turn a dial. Similarly, to adjust the TRVs (for goals 1 and 3), householders were required to turn a knob with a scale ranging from 1 to 5. For goal 2, the boost button was a simple 'on/off' button positioned on the programmer device. The pros and cons of these ways in which these controls facilitate appropriate execution are discussed further in the next section.

The stages so far make up the gulf of execution (Norman 1986). The fifth and sixth stages encompass the perception of the system and the users' interpretation of what they perceive, in relation to their goal. In the design model represented in Figure 6.4, elements of the heating controls that feedback the state of the system are shown with a thin border. The programmer had a red indicator light that illuminated during the set time periods where the boiler can come on if the thermostat has 'called for heat' (but will remain off, if the thermostat senses the set room temperature has been reached). This is clearly relevant for goals 1 and 3, as is feedback from the thermostat device, which indicates with a flame signal when calling for heat (but does not indicate that the boiler will only come on during scheduled time periods). For all three goals, boiler activation is a key variable. When active, the boiler illuminated an indicator light and made a noise (though these also indicated hot water provision). For the benefit of goal 3, the thermostat also displayed current room temperature for the room in which it was located, and the set points of the TRVs in unused rooms could also be perceived.

Norman's (1986) seventh stage of activity captures when the user evaluates the outcome. This is achieved by comparing their interpretation of the perceived system state with the original goals (Norman 1986). In the home-heating context, the point at which the evaluation may take place could vary. If the householder was feeling cold in the morning and evening, they could evaluate if the system was operating as intended to achieve goal 1 (provide routine comfort needs) by checking the device feedback. Where the thermostat showed the room temperature value below the set temperature and the flame indicator symbol, the programmer indicator was illuminated, and the boiler could be heard and its indicator was illuminated, they could accurately conclude that the system was working to pursue their goal. If the indicators on these devices were not present but the room temperature display on the thermostat matched the set temperature throughout the day, the householder could evaluate that the heating system had achieved the intended state (fulfilling goals 1 and 2). Outside the scheduled programmer periods, if the householder felt cold and had perceived from the thermostat that the room temperature value was below the set temperature value, then they could evaluate the need to provide ad hoc heating (goal 3) and specify to execute the 'boost' button (goal 2) to achieve the desired system state. To evaluate part of goal 3, that residual heat is not wasted when the house is unoccupied, with the existing system, the user would need to perceive the programmer indicator unlit a set time before planning to leave the house. The final part of goal 3, that heat is not emitted into long-term unused rooms, would result from understanding that a low set point on the specific TRV in the unused room would prevent hot water from entering that radiator.

This section has proposed a 'design model', 'compatible model' and recommended system state of a UK home-heating system. How these allow the gulf of evaluation and execution to be bridged has been discussed. Norman (1986) proposed that the design of the system image influences whether a compatible model is reinforced to the user when they interact with the system (Figure 6.2). How the system image of the heating system in the case study could influence the formation of a compatible mental model will be discussed in the next section.

6.3 SYSTEM IMAGE OF HOME HEATING

Norman (1986) emphasized the importance of level when considering the gulf of evaluation and execution. In the home-heating context, the system image was considered at the system level by envisioning the household as an 'interface', as well as at the device level, by examining the human–device interface for key home-heating controls. Both levels are represented in Figure 6.8. At the system level, the consequences of the prominence, connection and distribution of devices will be described. At the device level, the cognitive and physical usability of individual control interfaces, both cognitive and physical, will be considered. How these variables have the potential to impact the development of a compatible user mental model (Figure 6.5), as well as influence specific stages of Norman's (1986) seven stages of activity, is reflected upon in this chapter.

6.3.1 HOME HEATING AT THE 'SYSTEM' LEVEL

At the system level, Figure 6.8 shows control devices are distributed across the house, with different levels of prominence. The TRVs are positioned at ankle level adjoined to each radiator in each room. The central thermostat is positioned at eye level in the hall, exposed to high levels of traffic by occupants in the house. The programmer

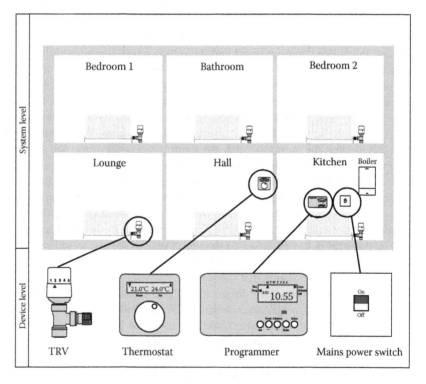

FIGURE 6.8   The 'system image' of the home-heating system, showing the layout and device interface of the home-heating elements from the 'compatible mental model'.

TABLE 6.2
Summary of Analysis of the System Image of the Heating System, with Possible Misunderstanding by the User

Control Device	Quick to Adjust	Easy to Adjust	Visually Prominent	Possible Misunderstandings (Level: S, System; D, Device)
Programmer	X	X	✓	Not connected to heating system (S,D)
				Independent device – not dependent on thermostat (S,D)
				Programmed times are the same as heat output times (D)
				Programmed times are the same as desired comfort periods (D)
House thermostat	✓	✓	✓	Not connected to heating system (S,D)
				Independent device – not dependent on programmer (S,D)
				Set temp knob ensures temperature of all rooms in house (D)
				Temp display measures temperature of all rooms in house (D)
				Varies temperature of boiler like variable valve control
Boost	✓	✓	X	Heat output from boiler is increased/amplified (D)
				Independent device – not dependent on thermostat (S,D)
				Not time-based control (D)
TRV	X	✓	X	Set point varies flow of hot water into the radiators (D)
				Fast-responding control (D)
				Not feedback-based control (D)
On/off switch	✓	✓	✓	Control for heating demand (D)
				Not connected to heating system (S,D)

and master power switch are just below eye level and positioned in the kitchen near the boiler. The boost button is a sub-feature of the programmer (Figure 6.8). The differing prominence of the control devices in Figure 6.8 are summarized in Table 6.2. This variation may play a part not only in which devices 'come to mind' as part of the user's mental model, influencing specification of the action sequence (stage 3), but which devices 'come to sight' when the user is perceiving and interpreting the state indicators on devices (stages 5 and 6). In Chapter 4, Revell and Stanton (2014) depict how information intake is filtered depending on the user's background experience. Devices that users do not expect to be part of the heating system may be treated as not relevant.

Figure 6.8 shows a lack of visible 'connections' between elements in the home-heating system. Whilst a simplified view of central heating at a system level,

Figure 6.8 reflects reality in that some devices transmit wirelessly, and pipes and wires that connect elements of the system are partially obscured by floorboards and walls, preventing continuous visual tracking from element to element. Cause and effect at a system level is not, therefore, explicit, which could prevent a compatible mental model being formed or reinforced. Johnson-Laird (1989) illustrates that where information is ambiguous, multiple UMMs can be formed. Variations in UMMs for a single system ultimately could lead to variations in the way users bridge Norman's (1986) seven stages of action (see Figure 6.3), resulting in differences found in users' behaviour with heating system as identified in the literature (e.g. Lutzenhiser and Bender 2008).

The distributed nature of heating system elements (Figure 6.8) also affects the ease with which users can gain feedback of the system state from control devices. To evaluate appropriate functioning of the system requires the user to relocate to the hall and kitchen to perceive and interpret indicator lights and icons on the thermo-stat, programmer and boiler (stages 5 and 6). To confirm the set point values on the control devices demands the user to be adjacent to the device, which in the case of TRVs are distributed across the household and require bending or crouching to read the value. For the user to accurately evaluate the system state from the various control devices requires focused attention and physical 'effort' from the user. Bainbridge (1992) proposes that 'meta-knowledge' (a form of data learned through experience) is incorporated with mental model data to determine the chosen strategy. The 'ease of evaluation' of the functioning of a system based on the various controls that need to be referred to for perception and interpretation (stages 5 and 6) may be stored as 'meta-knowledge'. This may result in 'short-cuts' or estimates from limited data being adopted by the user when performing their evaluation.

6.3.2   HOME HEATING AT THE DEVICE LEVEL

Figure 6.8 highlights the key control devices identified in the 'compatible mental model' (Figure 6.5). The interface of each control device may support, mislead or fail to reinforce the way devices function in the model in Figure 6.5. For example, the TRVs (Figure 6.8) are attached to each radiator, indicating custom control of the radiators within the system. The nature of the customization possible is misleading from the system image. The user is presented with a scale from 1 to 5 and set point adjustments are made by twisting a knob (Figure 6.8). There is no indication on the device of the relationship between the 1–5 scale and the temperature range at which hot water will be prevented from flowing into the radiator. Nor is there an indica-tion that the device responds to changes in room temperature, or that this is a slow-responding device. Crossman and Cooke (1974) suggest that manual operators of slow response systems need to be taught the control characteristics in order to secure the best results. The name of the device (thermostatic radiator valve) contains refer-ence to the 'feedback' function of the thermostat, as well as the control of fluid flow function of the valve which reflects well the device function. However, the device is not labelled by name, and the idea of a valve that users may have may relate to expe-rience with those that offer fast-acting variable obstruction (like a gas valve) rather than a slow-responding control that functions by allowing or blocking flow. The idea of variable flow is reinforced by the twisting motion of the knob, similar to that used on other variable flow devices (e.g. gas hob control, tap). Together, these elements seem to communicate to the user a 'natural mapping' (as described by Norman 2002) between the 1–5 scale and heat output, rather than the ability to set a temperature range for automatic flow control.

Like the TRVs, the central thermostat in the case study households (Figure 6.8) also requires a twisting motion to adjust its set point. Kempton's (1986) 'valve' folk model of thermostat function may similarly be encouraged by this twisting motion. This could therefore have an effect on the action specification and execution of home-heating goals (stages 3 and 4). On the thermostat interface, an LED display shows the chosen set point and the current room temperature (Figure 6.8). The room temperature label does not emphasize that this is the sole sample point used to calculate when to 'call for heat'. Whilst a 'flame' icon indicates when the thermostat calls for heat, there is nothing on the interface to communicate that boiler activation is also dependent on programmer-scheduled time periods. Both of these attributes of the interface are state indicators available for perception (stage 5) but could result in problems of interpretation (stage 6), and ultimately evaluation of the ability of the system to achieve desired home-heating goals (stage 7).

The programmer device (or timer) also does not identify its function by labels on the main casing. To set the scheduled time periods requires an involved series of steps and mode changes, obliging the support of a manual for first-time (or infrequent) adjustment. The process requires time periods to be inputted for each day of the week. The user inputs times into the device, prompted by labels such as 'on' and 'off'. The device does not make the distinction that these periods are possible 'boiler on' times, subject to the thermostat, rather than actual boiler on/off times. This fails to reinforce the conditional nature of boiler activation required for a compatible mental model (Figure 6.5). The programmer also does not distinguish to the user the difference between scheduled times and comfort times, taking into account residual heat. A time lag in heating up and residual heat are characteristics of a slow response system (Crossman and Cooke 1974). Whilst the program schedule is an automatic system, the automation relates to time, not consideration of thermodynamics within the home setting. The householder remains the manual controller of this aspect when programming the start and end times. Crossman and Cooke's (1974) analysis concludes that controllers need the necessary information about the behaviour of the system to make adequate decisions for control.

The boost button is located on the programmer, but the button name is misleading in terms of its function. The meaning of the word 'boost' in the United Kingdom is associated with ideas of increase, enhancement or amplification, providing a misleading metaphor (Lakoff and Johnson 1981). The user could reasonably believe more powerful heat flow, or higher temperatures will result, resulting in errors in the user's mental model. The term also fails to communicate effectively that its function is to provide 1 hour of ad hoc program schedule, which may influence decisions made at the action specification stage (stage 3). Like the program schedule, the boost button's dependency on the thermostat for boiler activation is not explicit. This may affect the compatibility of the resulting user's mental model, as well as prevent appropriate interpretation of the system state (stage 5).

The programmer also offers other features such as 'on', '24', 'advance' (see Figure 6.8), providing the user with a variety of options that may compete with the 'boost' feature at the action specification stage (stage 3). The boost feature is easy to activate, requiring only a single press without the need to choose a set point, assisting the execution stage (stage 4).

The final device highlighted in Figure 6.8 is the master power switch. This interface is a simple generic switch with on/off labels and red highlighting (meeting conventions for a power switch). Without awareness of the conventions for a mains power switch, users may believe this is a heating control, particularly if through experimentation its influence of the heating system is evident. This could result in inappropriate action specification and execution (stages 3 and 4). Conversely, without experimentation, there is nothing on the device to indicate its link to home heating, as opposed to other devices (e.g. cooker or oven). As a result, some users may not incorporate this in their home-heating mental model. Nevertheless, like the boost button, thermostat and TRV, it is easy to operate. Norman (1986:40) succinctly positions the effect of usability of input devices on the interactions people have with systems: 'Because some physical actions are more difficult than others, the choice of input devices can affect the selection of actions, which in turn affects how well the system matches with intentions'. Easy to operate devices, therefore, may be operated more readily than more appropriate 'less easy' to operate devices (such as the programmer). This may explain device selection that does not follow from the user's mental model description. Bainbridge's (1992) concept of 'meta-knowledge' further supports this sentiment. The 'ease of operation' of a device may be stored as 'meta-knowledge' that ultimately influences the behaviour specified and carried out (stages 3 and 4). Table 6.2 summarizes the user's experience of the system image of the home-heating interface, based on speed and ease of adjustment, prominence and possible misunderstandings. This will be used in the next section to help explain differences in the user mental models or reported behaviour, from those recommended by the expert.

6.4 THE USER'S MENTAL MODEL OF HOME HEATING – CASE STUDY RESULTS AND DISCUSSION

The previous sections have shown how Norman's (1986) seven stages of activity can apply to the home-heating system. A design model (see Figure 6.4) and proposed 'compatible' mental model of the system (Figure 6.5) that could help users to appropriately operate the system have been depicted. The control devices and settings required to achieve a 'system state' that could support assumed goals of householders with a young family have also been illustrated (Figure 6.6). The system image of the case study heating system (Figure 6.8) has been described at both the system and device levels and inferences have been made as to how this may influence the development of users' mental models, and where this may have an impact on Norman's (1986) seven stages of action.

In this section, how householders' mental model descriptions compare to the expert's idea of a 'compatible' mental model is shown. Householders' self-reported behaviour with home-heating controls is also compared with the recommended execution from Figure 6.6. With reference to the inferences of the effect of system image, these data will be analysed according to Norman's (1986) seven stages of action to highlight difficulties real users have in bridging the gaps of evaluation and execution with their home-heating system (Figure 6.3).

6.4.1   How Compatible Were the Case Study User Mental Models of Home Heating?

Norman (1983) argues the importance of understanding where users' mental models are erroneous and incomplete, and considers it the duty of designers to develop systems that aid users to develop coherent, usable mental models. To provide a visual comparison that illustrates how compatible the householders' user mental models were with the compatible model of home heating (Figure 6.5), elements that match those in Figure 6.5 have been represented in Figure 6.9 in black. Omitted elements from the compatible mental model are shown in light grey.

Figure 6.9 shows that householders from the case study varied in the key elements of the design model present in their mental model descriptions. The boiler, radiators and connecting pipes were evident in all, but the control devices, links and associated rules differed for each participant. The thermostat control was evident in all model descriptions except P1, though for P6, the room temperature display and the rule for 'calling for heat' were not depicted. The programmer device and schedule was evident in all but P3's model description, but the boost and advance buttons were not depicted by any participant. The TRV control was evident in half of the descriptions (P1, P4 and P6), but the slow response, conditional rule for turning on and off and the feedback link from room temperature were absent in all cases. Only participants 2, 4 and 5 depicted the conditional rule for the boiler, and only participants 3 and 5 depicted the master on/off switch. Half of the participants (P1, P3 and P6) did not specify that the heating settings were designed to influence 'room' temperature (rather than whole house temperature, or body temperature). Figure 6.10 summarizes which elements of householders' user mental models match (to the right, in grey), or are missing (to the left, in black) from the expert-recommended 'compatible' model (Figure 6.5). This table provides an indication of where the user interface of the home-heating system in this study promotes, or fails to promote, an appropriate user mental model.

6.4.2   How Appropriate Were Case Study Self-Reported Behaviour of Home-Heating Operation?

To understand how householders specify and execute their goals with their home-heating systems, Figure 6.11 compares their self-reported behaviour. The x-axis displays time divided by days of the week, with shading representing night (10 pm to 6 am). The left y-axis measures the thermostat set point temperature that corresponds to the dashed line graph. The right y-axis displays a series of scales (on/off or 5-point scale), so multiple devices and their set points can be shown on one graph. The programmer schedule on/off times are displayed at the base with a solid graph, followed by boost/advance, programmer on/off button (in bold line). Towards the top, the TRV set points are displayed, and where different radiators have different set points, these are labelled.

From Figure 6.11 we can see participants 1, 4, 5 and 6 report that they use the programmer device to schedule routine periods of operation, as recommended by the expert. P5, however, also reports ad hoc adjustments to the schedule to remove

P1 'Key compatiple elements to the design model evident in mental model description' (missing elements greyed out)

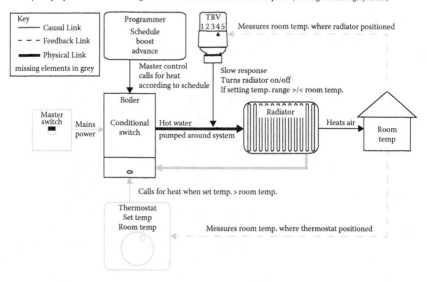

P2 'Key compatiple elements to the design model evident in mental model description'

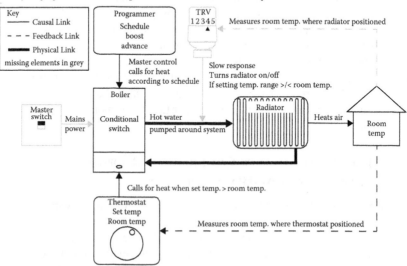

FIGURE 6.9   Key elements compatible to design model, for each participant – greyed-out areas are missing elements.   (*Continued*)

a time period when the house will be unoccupied. All participants, except P1, report use of the thermostat with participants 4, 5 and 6, keeping this at a constant setting. For participants 2 and 3, the thermostat is reported as the sole device that they use, with routine changes in set points, as well as 'ad hoc' changes (Figure 6.11). None of the participants report the use of the boost/advance button, as recommended by the expert for 'ad hoc' changes to provide heating on demand between scheduled times.

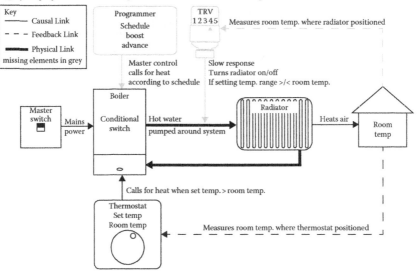

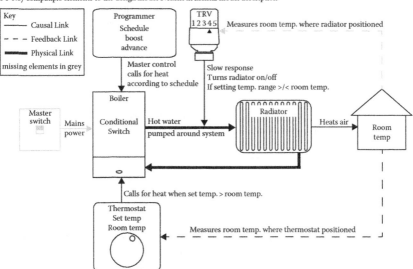

FIGURE 6.9 (*Continued*)   Key elements compatible to design model, for each participant – greyed-out areas are missing elements.   (*Continued*)

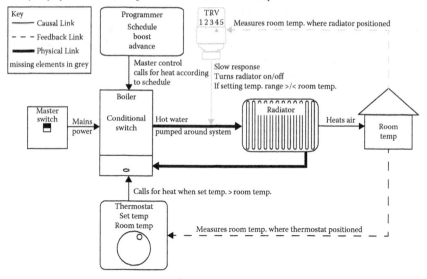

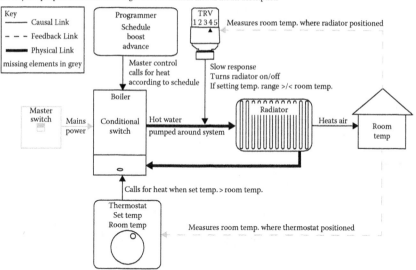

FIGURE 6.9 (*Continued*)   Key elements compatible to design model, for each participant –
greyed-out areas are missing elements.

Participants 1 and 6 report the use of the programmer on/off button (absent from the
'compatible mental model') in order to fulfil this goal. Three participants report setting their TRVs: participants 1 and 6 use them as recommended by the expert, keeping them at a constant setting, and using a lower setting for rooms where heat needs
to be limited. Participant 4 reports adjustment to all radiators in the household on a

Required element	Connection/Function	Participants with Missing Elements						Participants with Compatible Elements					
Master switch	Connected to → Boiler		P6	P4	P2	P1		P3	P5				
	Electrical isolation	P6	P4	P3	P2	P1		P5					
Boiler	Connected to → Radiators							P1	P2	P3	P4	P5	P6
	Pumps hot water around system							P1	P2	P3	P4	P5	P6
	Conditional switch			P6	P3	P1		P2	P4	P5			
Radiators	Connected to → Room temp			P6	P3	P1							
	Heats air							P1	P2	P3	P4	P5	P6
Room temp	Connected to → TRV	P6	P5	P4	P3	P2	P1						
	Connected to → Thermostat				P6	P1		P2	P3	P4	P5		
	Provides sample of temp.value				P6	P1		P2	P3	P4	P5		
Thermostat	Connected to → Boiler					P1		P2	P3	P4	P5	P6	
	Temp set point					P1		P2	P3	P4	P5	P6	
	Room temp display				P6	P1		P2	P3	P4	P5		
	Calls for heat IF room temp < set temp				P6	P1		P2	P3	P4	P5		
Programmer	Connected to → Boiler					P3		P1	P2	P4	P5	P6	
	Schedule times					P3		P1	P2	P4	P5	P6	
	Calls for heat during timeslots					P3		P1	P2	P4	P5	P6	
	Boost – add 1 hour timeslot	P6	P5	P4	P3	P2	P1						
	Advance to next timeslot state	P6	P5	P4	P3	P2	P1						
TRV	Connected to → Pipes			P5	P3	P2		P1	P4	P6			
	Set points			P5	P3	P2		P1	P4	P6			
	Blocks or allows hot water to radiator			P5	P3	P2		P1	P4	P6			
	Allows access IF room temp < set point	P6	P5	P4	P3	P2	P1						
	Slow response	P6	P5	P4	P3	P2	P1						

FIGURE 6.10 Figure to compare the elements of householders' user mental models with those in the proposed 'compatible' model.

daily basis, with heat being limited from all radiators during the day, all radiators set to 3 during the evening and some radiators set back to 1 during the night.

Figure 6.10 was constructed with reference to householders' transcripts and Figure 6.11, to understand how householders bridge stages 3 and 4 (action specification and execution) of the gulf of action, and provide a comparison to that recommended by

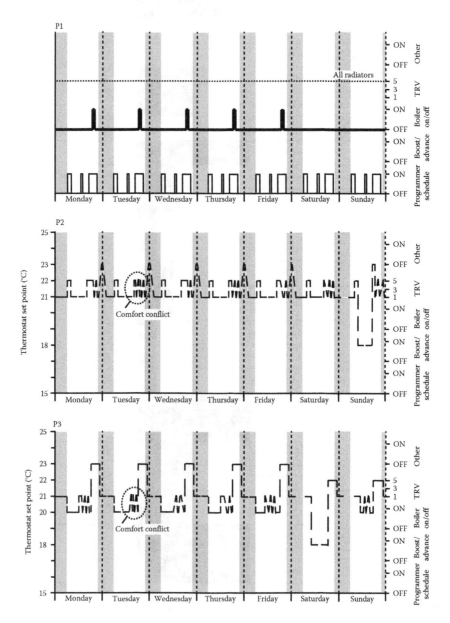

FIGURE 6.11 Householders' self-report of typical use of home-heating controls over a week period.   *(Continued)*

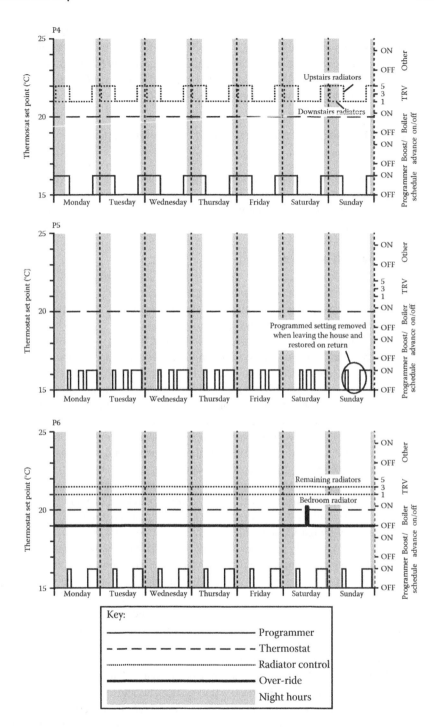

FIGURE 6.11 (Continued) Householders' self-report of typical use of home-heating controls over a week period.

the expert. For each goal, the expert recommended control device and actions specifications are shown on the left. Where householders agree with the recommendations is marked with a tick. Where the recommendations are partially followed, or an alternate cause of action is described, the amendment is shown in text. Recommendations that were not followed are marked with an 'X' (Table 6.3).

6.4.3 A Discussion of the Seven Stages of Activity When Users Operate Their Home-Heating System

The aim of this section is to evaluate case study data presented in the previous section within the framework of Norman's (1986) seven stages of action. Reference will be made to the analysis of the existing interface to explain the user mental models and execution of heating controls.

6.4.3.1 The Gulf of Execution

Norman's (1986) first two stages of activity are setting goals and intentions. Assumptions were made about the goals for this audience group. To check that user goals were in line with the assumptions made by the expert, the transcripts from QuACk were analysed to identify and categorize householders' goals. Whilst a larger range of goals were identified, the three goals assumed by the expert were evident in the case study group. The outputs in Figure 6.8 were checked against the transcripts to ensure they reflected actions associated with these three goals and confirm that meaningful comparison could be made with the expert-recommended actions (see Figure 6.6).The variations in the control devices depicted in Figure 6.8 shows the participants varied in their intentions when pursuing their heating goals.

Table 6.3 represents how users specify actions (stage 3) and execution actions (stage 4) when bridging Norman's (1986) gulf of evaluation and execution. Table 6.3 shows the householders in this case study were in most agreement with the expert recommendations when specifying actions to fulfil goal 1 (meeting routine comfort needs). Four of the participants used the program schedule to provide heating that met their routine comfort needs. Figure 6.11 shows that the time and number of repeated slots vary by participant, but, with the exception of P4 (who chose a single overnight time slot), closely reflect comfort periods assumed by the expert. All these participants had the programmer in their mental model and understood the function of the programmer (see Figures 6.9 and 6.10). Participants 2 and 3 did not use the programmer at all. Participant 2's action specification is explained by the programmer being absent from their mental model (Figures 6.9 and 6.10), which is surprising given its prominent location. The lack of functioning labelling at the device level of the system image, and no visual connection to other devices at the system level, may be responsible for this device not being considered part of the heating system (Table 6.2). Participant 2's mental model showed an appropriate understanding of the function of the programmer (Figures 6.9 and 6.10). The transcript revealed that his choice not to use this control device was a usability issue. He described the device as 'too much hassle' to operate, providing support to Norman's (1986) view of the importance of usability in the selection of actions.

TABLE 6.3

Comparison of Householders' Actions to Achieve Goals, with Expert Recommendations

Goal	Expert	P1	P2	P3	P4	P5	P6
1 – Meet routine comfort needs	*Programmer*						
	Schedule used	✓	X	X	✓	✓	✓
	Thermostat						
	Constant set point (18°C–20°C)	X	Routine adjustment	Routine adjustment	✓	✓	✓
	TRV						
	Constant set point (set to 3)	✓	✓	✓	Routine adjustment	X	✓
2 – Ad hoc heating	*Boost*						
	Turn on when needed	X	X	X	X	X	X
	Alternate where option absent from user mental model	*Override* Turn on when cold for an hour	*Thermostat* Turn up when cold	*Thermostat* Turn up when cold	X	*Schedule* Add time slot	*Override* Turn on when cold for an hour

(Continued)

TABLE 6.3 (*Continued*)

Comparison of Householders' Actions to Achieve Goals, with Expert Recommendations

Goal	Expert	P1	P2	P3	P4	P5	P6
3 – Avoid wasting heat	*Programmer*						
	Time slot ends earlier than comfort slots	Time slot matches comfort slots ✓	X	X	Time slot matches comfort slots ✓	Remove timeslot ✓	Time slot matches comfort slots ✓
	TRV						
	Reduce set point for rooms unused long-term	X	X	X	For rooms unused short-term ✓	X	✓
	Temp display						
	Check before using override	X	X	X	X	X	✓
	Alternate where option absent from user mental model	n/a	*Thermostat* Turn down when house unoccupied	*Thermostat* Turn down when house unoccupied	n/a	n/a	n/a

To support goal 1, Table 6.3 shows that the expert specification to keep the thermostat at a constant setting between 18°C and 20°C was matched by participants 4, 5 and 6. Participants 4 and 5 held an appropriate concept of the thermostat and its function (Figures 6.9 and 6.10), but participant 6 chose an appropriate specification despite lacking an understanding of the role of the thermostat. In Chapter 4, Revell and Stanton (2014) revealed that their full user mental model promoted appropriate action through an alternately compatible model of the system. Participant 1 did not include the thermostat in their user mental model nor the action specification (Figures 6.9 and 6.10). Again, this omission is surprising given the prominence of the device (Table 6.2), but the lack of function labelling and visual 'connection' to other heating elements in the system image could provide the explanation. This lack of awareness provides a positive unintended consequence, as it would ensure a 'constant' setting for the thermostat (though not necessarily at the appropriate set point). For participants 1 and 6, appropriate action with the thermostat could therefore arise from elements in their user mental models that differ from the 'compatible mental model' (Figure 6.5). Table 6.3 and Figure 6.9 show that participants 2 and 3 stand out by using the thermostat as their sole control. For goal 1, routine adjustments of the thermostat result in deliberate increase and decreases in set point as comfort needs change throughout the day. This action specification is explained by these participants choosing not to use the programmer schedule. Used as a sole device, routine adjustments of the set point would be an appropriate way to achieve goal 1 and is expected from users holding a 'feedback' mental model of home heating as defined by Kempton (1986). Figures 6.9 and 6.10 show both participants had a compatible understanding of the thermostat. This illustrates how recommendations for home-heating controls, which assume a control is used in conjunction with another control (e.g. thermostat and programmer), would need amending for users who were using one of these controls in isolation.

The TRV was the final control recommended by the expert to achieve goal 1. A consistent mid-range setting was suggested, which was adopted from participants 1 and 6 (see Figure 6.11, Table 6.3) despite a basic understanding of the way the device functioned (Figures 6.9 and 6.10). The lack of action may have reflected the effort (Table 6.2) required to change control devices in multiple locations. Participant 4 also had an equivalent understanding of the TRV, which omitted its role as a slow-responding thermostat. This participant specified a labour-intensive execution that required him to routinely adjust the TRVs on a twice daily basis so they were set low and moderate in unused and used rooms, respectively. This approach would be appropriate for a fast-responding control, but for the TRVs installed in this household, the response time would result in occupants spending time in rooms that were uncomfortably cold, and leaving unoccupied rooms at a comfortable temperature, whilst the change in setting took time to take effect. The system image for the TRV appears similar to fast-responding valves which may have contributed to this behaviour (Table 6.2). Participants 2, 3 and 5, did not report using the TRV to achieve their heating goals, nor had them as part of their user mental models. This absence may have been due to their low prominence due to the ankle-level position (Table 6.2). Again, this omission implies these devices are kept at a constant default setting, unintentionally complying with recommended advice.

For goal 2 (ad hoc heating between scheduled time), the action specification rec-ommended by the expert was to press the 'boost' button. None of the householders had the boost feature in their user mental models (Figures 6.9 and 6.10), and this strategy was not evident in their behaviour (Figure 6.11, Table 6.3). This omission may have been due to lower prominence as a 'sub' feature of the programmer, or the ambiguous label may have confused users, so they hesitated to operate this fea-ture (Table 6.2). Participants instead chose to operate the manual override on the programmer (P1 and P6), increase the thermostat control on demand (P2 and P3) or access the programmer schedule to add a time slot (P5). These strategies do fulfil goal 1, but may jeopardize goal 3, as they risk the user forgetting to reset the device to the original setting. This could waste energy by providing heat for longer or extra periods than required to meet comfort needs. The consequence of overheating by thermostat control could result in 'comfort battles', with frequent adjustments of the thermostat in opposing directions (Kempton 1986), as seen in the output of P2 and P3 (Figure 6.11).

The action specification provided by the expert to achieve goal 3 (avoid wasting money on heating when not required) was to: (1) end scheduled time slots earlier than comfort slots; (2) reduce TRV set-point for long-term unused rooms and (3) check the temperature display before deciding to execute 'ad hoc' heating. Whilst four of the participants used the programmer schedule, which contributes to this goal, their chosen 'off' time matched (rather than preceded) the end time of their comfort needs. This implies wasted energy where the home retains a comfortable heat level when this is not required. This applied even for participants whose under-standing of the program schedule was compatible with the design model (P4, P5 and P6; see Figures 6.9 and 6.10). Participants 2 and 3, who used the thermostat as the sole control, fulfilled goal 3 by turning the thermostat device down below comfort levels, when away for the day, but also made these changes at the time of leaving the house, rather than a period of time before, so unused residual heat is still wasted. The action specification, of setting an earlier end time for heat output requires the user to incorporate an understanding of thermodynamics, with the functioning of the heating system. Compatibility with the design model alone is therefore insufficient, and the system interface at the device level for the programmer fails to highlight the difference between heat output times and comfort levels periods (reinforcing the view of Crossman and Cooke (1974) that operators of slow response systems need more state information or training to be able to make appropriate settings or adjust-ments. Participant 5 also described another strategy, by actively removing time slots from the programmer when going out for the day (Figure 6.11). This is an effec-tive and appropriate way of preventing wasted energy that conforms to a compatible underlying mental model of the system. However, there is a risk of forgetting to reset the routine settings that would result in comfort goals not being met. Strategies by participants 2, 3 and 5 all carry the risk that householders will forget to execute the initial action, allowing energy to be wasted.

Participant 6 adhered to the recommended action specification with TRVs, by adjusting rooms not requiring heating long-term, to a low setting and so saving energy (Figure 6.11, Table 6.3). Missing compatible elements in her mental model

(Figures 6.9 and 6.10) did not prevent specification of this action. Participant 4, whose compatible elements of the TRV were identical to that of participant 6 (see Figures 6.9 and 6.10), intended to support goal 3 by minimizing heat during the day and in unused rooms throughout the evening and night. The action specification used required routine daily adjustments, inappropriate for a slow response control (Figure 6.11, Table 6.3). This illustrates that the incomplete mental model of the TRV is not sufficient to ensure appropriate action, particularly as the system image at the device level may encourage the idea of a fast-responding control (Table 6.2).

Participant 6 was also the only householder who stated (in transcripts) checking the temperature display on the thermostat device before executing ad hoc heating. This is surprising since participant 6's user mental model was missing the temperature display element, link to room temperature and function of thermostat. Again this reveals that appropriate behaviour can derive from a model that is not compatible with the design model in the way suggested by the expert (see Revell and Stanton 2014, for more in-depth analysis of this participant).

6.4.3.2 The Gulf of Evaluation

So far, how user mental models and the system interface have influenced householders' activities to bridge Norman's (1986) gulf of execution has been discussed. Norman's (1986) final three stages of activity bridge the gulf of evaluation, and involve perception, interpretation and evaluation (Figure 6.1). The system state is an information input for these stages (Figure 6.11); therefore, how the system image provides feedback to the user about the system state is of particular importance in this analysis. Whilst manufacturers provide indicators on their devices to inform the user of the system state, the user may give greater importance to other forms of perceptible feedback when interpreting the system state. This may affect the appropriateness of their evaluations when determining whether the system has achieved their goals. Table 6.4 shows the feedback cues evident on the householders' mental model descriptions. Householders were not specifically asked which feedback they took onboard when trying to achieve different goals, so this analysis will be more general.

Table 6.4 shows that the feedback cues recommended by the expert for evaluating the state of the system appear sporadically in the householders' mental model descriptions. Participant 5 uses most of these cues, including the thermostat, programmer and boiler indicators in his model description, and is the only participant who would be able to evaluate the system state in the same way suggested by the expert. His action specification in Figure 6.11 shows an appropriate use of devices. Participant 6 is the only one whose model description included the temperature display on the thermostat, but does not mention any of the other device indicators. Participants 2 and 5 produced model descriptions that included the boiler light, and participants 1 and 3 had model descriptions containing the programmer light and thermostat indicator, respectively. Due to the minimal cues present, these participants would need to make assumptions about the system or use other cues to form their evaluation. Table 6.4 shows that the warmth of the radiator is a perceptual cue that is used by all, but participant 3, as feedback that the system is operating. This is not a reliable cue to

TABLE 6.4

Summary of Perceptual Cues Used by Participants to Evaluate the State of the System

			Feedback Cue				
Participant	Thermostat Indicator	Programmer Light	Boiler Light	Temp Display	Boiler Sound	Radiator Warmth	Other
P1		✓				✓	Body warmth
P2			✓			✓	
P3	✓				✓		Warmth of house
P4			✓		✓[a]	✓	Warmth of Pipes
P5	✓	✓	✓			✓	
P6					✓	✓[a]	✓

[a] Participants aware that sound also indicates hot water.

determine if the action specification has been successful. A warm radiator does not necessarily imply that the boiler is active, or that the TRV is open (there may be residual heat). An already warm radiator can only inform the interpretation that hot water is currently in the radiator. A warming up radiator, on the other hand, can be used to indicate that the TRV is open, and the boiler is active. It cannot accurately enable the evaluation that a set point value on the thermostat has been achieved, however. This means goal 1 is difficult to evaluate effectively with this cue. A cold radiator also does not automatically indicate that the programmer and thermostat are not operating, or that the TRV is closed. The room temperature in the hall may be at, or above the thermostat set point, preventing the boiler being activated, and water being sent around the system. This, therefore, is not a definitive way of determining that energy will not be wasted on leaving the house (goal 3). To evaluate if goal 2 (the need for ad hoc heating) is required, a cold radiator could be used to evaluate that the boiler is not currently active only in rooms where the user knows the TRV setting is open. Further perceptual cues will need to be sought to determine the appropriate action specification for activating the boiler (e.g. thermostat set point, programmer status). Misunderstanding the meaning of the radiator temperature could result in users believing the heating is off when they leave a house unoccupied or assuming the heating has not achieved the temperature set by the thermostat, resulting in unnecessary override of the system. Both of these misunderstandings could result in wasted energy, jeopardizing goal 3. Inappropriate override of the system could also result from the inappropriate interpretation of other cues. Participants 1 and 3 indicated on their mental model description that they could tell the heating system was working based on the warmth of the whole house or their own body comfort level. Due to the effect of thermodynamics on how well different rooms in the house are heated, and the effect of activity levels, dress and comfort preferences on body comfort level, both of these are poor ways for determining if the desired system state has been achieved. In the case of participant 3, this resulted in the thermostat set points being increased or decreased based on the felt temperature of different rooms in the house, rather than the desired set temperature (Figure 6.11, Table 6.3). Participant 1 also reported overriding a system during a scheduled time period when feeling cold and evaluating the heating had not achieved the set goal (Figure 6.11, Table 6.3). A full analysis of these participants' responses is described in Chapter 4 and Revell and Stanton (2014). Participant 6 is the only participant who refers to the temperature display before considering ad hoc heating, enabling her to evaluate appropriately if the system state has been reached. Despite not describing the other recommended cues, her operation fits the recommended pattern (Figure 6.11, Table 6.3) suggesting interpretation of this type of feedback is particularly useful for evaluating if goals have been obtained.

6.5   CONCLUSIONS

This chapter set out to apply Norman's (1986) gulf of evaluation and execution to the domestic heating system. It took a different approach to previous research, by examining the 'scaffolding' that underlies the bridges over these gulfs. User mental models from householders in a case study were used to inform analysis at each of Norman's (1986) seven stages of activity. The analysis used a 'standard' derived from expert-recommended examples of a 'compatible mental model' and 'action specification' to evaluate the appropriateness of the way householders think and behave with their home-heating system.

The purpose of this analysis was to gain insights into why householders use their heating system inappropriately, often leading to wasted energy. These insights were thought to be important if considering the design of digital interfaces tasked with enabling users to better control their home-heating systems. It is concluded that this approach was useful at identifying issues that hinder appropriate operation of home heating. As an aid to those considering the design of home-heating interface systems, Table 6.5 summarizes how characteristics of the system image of the home-heating system in our case study influenced the user mental model, with examples taken from the case study data as to how for the three proposed home-heating goals, this 'scaffolding' had an impact on Norman's seven stages of activity.

Key findings that arose from analysis from the perspective of mental models were: (1) users do not necessarily perceive all the available key home-heating components in their home, (2) users do not understand how all the key heating components are connected and (3) users can misunderstand how key heating components function. These issues all contribute to poorer performance in terms of 'appropriate consumption', meaning energy is wasted, or frustration is caused as goals are not achieved. It was found, however, that a compatible mental model of how the heating system operates is not sufficient for the user to effectively achieve their heating goals. To ensure comfort and avoid wasting energy, expert advice recommended assigning end times for heat output at an earlier time than the end of a comfort period (so residual heat does not go unused). Also, for outlying rooms far from the thermostat, temperature

TABLE 6.5

How the System Image of the Home-Heating System Can Effect User Mental Models That Underpin Norman's (1986) Seven Stages of Action

Level	Characteristics of Existing UI	Impact of Existing UI on UMMs	Effect on Goal 1: Maintain Routine Comfort Needs (Numbering Relates to Norman's (1986) Stage of Activity)	Effect on Goal 2: Ad hoc Heating (Numbering Relates to Norman's (1986) Stage of Activity)	Effect on Goal 3: Avoid Wasting Heat (Numbering Relates to Norman's (1986) Stage of Activity)
System level system	Thermodynamics data missing	Thermodynamics missing from UMM	*Set long-term controls:* *a. Program schedule* *b. Thermostat set point* *c. TRV set point* 5 – Cannot perceive how thermodynamics effect the speed and distribution of heat around the home 6 – May interpret comfort levels or temp display before thermodynamic variable has had time to fulfil comfort goals 7 – May evaluate that long-term settings do not meet comfort goals and conclude that target temp/schedule durations need to increase	*Override heating system:* *a. 1 – Press boost button* n/a	*Utilize residual heat* *a. 1 – Schedule early end times* *Avoid heating used rooms* *b. 2 – Set TRV to low* *Check comfort level* *c. 3 – Compare to room temp* 2 – Cannot form intention to use residual heat 3 – Will not specify early end times for schedule 4 – Will not program early end times 5 – Cannot perceive how residual heat is being used 6 – Cannot interpret how much residual heat provides benefit to occupants 7 – Cannot evaluate if residual heat has been used

(Continued)

TABLE 6.5 (Continued)

How the System Image of the Home-Heating System Can Effect User Mental Models That Underpin Norman's (1986) Seven Stages of Action

Level	Characteristics of Existing UI	Impact of Existing UI on UMMs	Effect on Goal 1: Maintain Routine Comfort Needs (Numbering Relates to Norman's (1986) Stage of Activity)	Effect on Goal 2: Ad hoc Heating (Numbering Relates to Norman's (1986) Stage of Activity)	Effect on Goal 3: Avoid Wasting Heat (Numbering Relates to Norman's (1986) Stage of Activity)
System level system cont.	Range of controls	• Multiple control devices present in UMM • Hierarchy of controls may not be evident	3 – May form less appropriate specification if importance of programmer and thermostat not understood 4 – May set up long-term controls with inappropriate set points 5 – May give inappropriate weighting to particular state indicators (e.g. programmer on indicator, rather than boiler on indicator) 6 – May misinterpret state of system (e.g. assume boiler should be active when any control indicator is showing) 7 – May evaluate that system is operating ineffectively when it is responding appropriately	3 – May select an alternate specification to override heating 4 – May override heating execution with less appropriate (e.g. thermostat increase/boiler on-off button)	3 – May form less appropriate specification to avoid wasting heat (e.g. adjusting TRVs in used rooms) 4 – May execute controls inappropriately (e.g. frequent adjustment of TRVs) 5 – May give inappropriate weighting to particular state indicators (e.g. TRV set points rather than schedule end times) 6 – May inappropriately interpret a set point as an indicator of energy being saved (e.g. TRV just after turning down) 7 – May overestimate the reduction in wasted energy due to less appropriate settings

(Continued)

TABLE 6.5 (Continued)

How the System Image of the Home-Heating System Can Effect User Mental Models That Underpin Norman's (1986) Seven Stages of Action

Level	Characteristics of Existing UI	Impact of Existing UI on UMMs	Effect on Goal 1: Maintain Routine Comfort Needs (Numbering Relates to Norman's (1986) Stage of Activity)	Effect on Goal 2: Ad hoc Heating (Numbering Relates to Norman's (1986) Stage of Activity)	Effect on Goal 3: Avoid Wasting Heat (Numbering Relates to Norman's (1986) Stage of Activity)
System level system cont.	Lack of visible connection between devices	• Some connections between components missing in UMMs • Interdependency of devices not appreciated	3 – Inappropriate specification may result if dependency between programmer and thermostat is not understood 4 – User may fail to make adjustments to both schedule and thermostat 5 – May perceive program schedule activation as independent of the thermostat set point for turning on boiler 6 – May interpret programmer as not functioning if thermostat has stopped boiler 7 – May evaluate long-term settings have not been made	5 – May perceive boost status, as independent of the thermostat set point for turning on boiler 6 – May interpret boost as not functioning if thermostat has stopped boiler 7 – May evaluate heating schedule override has not been achieved	3 – Missing feedback loops from room temp to TRV may result in inappropriate specification 4 – User may execute TRV controls inappropriately (e.g. frequent adjustment to save energy) 5 – User may focus perception on TRV set point and overlook more important state indicators for saving energy 6 – User may interpret low TRV set points as indication that energy is not being emitted 7 – User may overestimate energy saved by adjusting TRVs

(Continued)

TABLE 6.5 (Continued)

How the System Image of the Home-Heating System Can Effect User Mental Models That Underpin Norman's (1986) Seven Stages of Action

Level	Characteristics of Existing UI	Impact of Existing UI on UMMs	Effect on Goal 1: Maintain Routine Comfort Needs (Numbering Relates to Norman's (1986) Stage of Activity)	Effect on Goal 2: Ad hoc Heating (Numbering Relates to Norman's (1986) Stage of Activity)	Effect on Goal 3: Avoid Wasting Heat (Numbering Relates to Norman's (1986) Stage of Activity)
System level system cont.	Variable prominence of devices	Some control devices missing from UMMs	3 – Cannot form recommended specification if programmer, thermostat or TRV missing from UMM 4 – Cannot set up as recommended 5 – May not perceive state indicators on control displays absent from UMM (e.g. thermostat calling for heat/ programmer on times, set points of TRVs) 6 – May make interpretations from incomplete state indicators (e.g. assume boiler should be on due to schedule, when thermostat has turned off when set temp achieved)	3 – Cannot form recommended specification if boost control missing from UMM 4 – May override heating with less appropriate execution (e.g. thermostat increase/boiler on-off button) 5 – May not perceive state indicators on control displays absent from UMM (e.g. thermostat calling for heat)	3 – Cannot form recommended specification if programmer, temp display or TRV missing from UMM 4 – Cannot set up as recommended 5 – Cannot compare comfort to room temp if temp display missing from UMM 6 – May misinterpret feeling cold is due to heating system rather than low activity 7 – May evaluate heating system should be overridden unnecessarily

(Continued)

TABLE 6.5 (*Continued*)

How the System Image of the Home-Heating System Can Effect User Mental Models That Underpin Norman's (1986) Seven Stages of Action

Level	Characteristics of Existing UI	Impact of Existing UI on UMMs	Effect on Goal 1: Maintain Routine Comfort Needs (Numbering Relates to Norman's (1986) Stage of Activity)	Effect on Goal 2: Ad hoc Heating (Numbering Relates to Norman's (1986) Stage of Activity)	Effect on Goal 3: Avoid Wasting Heat (Numbering Relates to Norman's (1986) Stage of Activity)
System level system cont.			7 – May evaluate incorrectly that heating system is not functioning as it should	6 – May make interpretations from incomplete state indicators (e.g. assume boiler should be on due to boost, when thermostat has turned off when set temp achieved) 7 – May evaluate incorrectly that heating system is not functioning as it should	
	Distributed control devices	Meta-data associated with UMM components include effort for access/adjustment	3 – TRV adjustments not specified as involve physical effort 4 – TRVs not adjusted (though default setting may be appropriate)	n/a	3 – TRV adjustments not specified as involve physical effort 4 – TRVs not adjusted

(Continued)

TABLE 6.5 (Continued)

How the System Image of the Home-Heating System Can Effect User Mental Models That Underpin Norman's (1986) Seven Stages of Action

Level	Characteristics of Existing UI	Impact of Existing UI on UMMs	Effect on Goal 1: Maintain Routine Comfort Needs (Numbering Relates to Norman's (1986) Stage of Activity)	Effect on Goal 2: Ad hoc Heating (Numbering Relates to Norman's (1986) Stage of Activity)	Effect on Goal 3: Avoid Wasting Heat (Numbering Relates to Norman's (1986) Stage of Activity)
Device level – Thermostat	Feedback function not communicated Temp sample unclear	• Feedback link missing • Function of device considered 'valve' • UMM assume whole house/comfort sampled	3 – Thermostat not specified as a long-term control 4 – Adjustments made to the set point to fit routine comfort needs (rather than single setting chosen) 3 – Specify thermostat adjustment for outlying rooms/comfort 4 – Inappropriate adjustments made to thermostat set point 5 – Temp value perceived as whole house/comfort 6 – Value that does not match users' physical experience in outlying rooms/own comfort may result in confusion 7 – User may evaluate that display is faulty/system is not working properly	n/a n/a	3 – Thermostat may be included in energy-saving specification inappropriately 4 – Unnecessary adjustments made to thermostat 6 – Value that does not match users' physical experience in outlying rooms/own comfort may result in confusion 7 – User may evaluate that display is faulty/system is not working properly

(Continued)

TABLE 6.5 (Continued)

How the System Image of the Home-Heating System Can Effect User Mental Models That Underpin Norman's (1986) Seven Stages of Action

Level	Characteristics of Existing UI	Impact of Existing UI on UMMs	Effect on Goal 1: Maintain Routine Comfort Needs (Numbering Relates to Norman's (1986) Stage of Activity)	Effect on Goal 2: Ad hoc Heating (Numbering Relates to Norman's (1986) Stage of Activity)	Effect on Goal 3: Avoid Wasting Heat (Numbering Relates to Norman's (1986) Stage of Activity)
Device level – Programmer	Conditional link to thermostat not communicated	• Conditional rule for boiler operation missing	5 – Schedule on indicators may be misperceived as 'boiler on' periods 6 – User may be confused if boiler is off during scheduled periods 7 – User may evaluate that long-term settings are not effective		2 – Cannot form intention to use residual heat
	Consideration of residual heat on schedule end times not emphasised	• Thermodynamics variable missing from UMM	n/a	n/a	3 – Will not specify early end times for schedule 4 – Will not program early end times
	Complicated set up procedure for schedule	Metadata associated with UMM components include effort for access/adjustment	3 – Programmer omitted from specification 4 – On/off times not scheduled	n/a	3 – Programmer omitted from specification 4 – On/off times not scheduled

(Continued)

TABLE 6.5 (Continued)

How the System Image of the Home-Heating System Can Effect User Mental Models That Underpin Norman's (1986) Seven Stages of Action

Level	Characteristics of Existing UI	Impact of Existing UI on UMMs	Effect on Goal 1: Maintain Routine Comfort Needs (Numbering Relates to Norman's (1986) Stage of Activity)	Effect on Goal 2: Ad hoc Heating (Numbering Relates to Norman's (1986) Stage of Activity)	Effect on Goal 3: Avoid Wasting Heat (Numbering Relates to Norman's (1986) Stage of Activity)
Device level – Boost	Conditional link to Thermostat not communicated	• Conditional rule for boiler operation not applied to boost operation	n/a	5 – Boost activation perceived to indicate boiler activation 6 – If boiler not active when boost is on, use may be confused 7 – User may evaluate that override was unsuccessful	n/a
	Function of device not communicated effectively (e.g. misleading name)	• Function misunderstood	n/a	3 – User may specify alternate controls for override 4 – User may execute alternate controls for override	n/a

(Continued)

TABLE 6.5 (Continued)

How the System Image of the Home-Heating System Can Effect User Mental Models That Underpin Norman's (1986) Seven Stages of Action

Level	Characteristics of Existing UI	Impact of Existing UI on UMMs	Effect on Goal 1: Maintain Routine Comfort Needs (Numbering Relates to Norman's (1986) Stage of Activity)	Effect on Goal 2: Ad hoc Heating (Numbering Relates to Norman's (1986) Stage of Activity)	Effect on Goal 3: Avoid Wasting Heat (Numbering Relates to Norman's (1986) Stage of Activity)
Device level – TRV	Feedback function not communicated	• Feedback link missing • Function of device considered 'valve'	3 – Missing feedback loops may result in inappropriate specification 4 – User may execute TRV controls inappropriately (e.g. frequent adjustment to provide comfort) 5 – User may focus perception on TRV set point at expense of schedule times/thermostat set point 6 – User may interpret TRV high setting to indicate higher volume/temp heat is emitted from radiators saved energy 7 – User may overestimate contribution to comfort levels	n/a	3 – Missing feedback loops may result in inappropriate specification 4 – User may execute TRV controls inappropriately (e.g. frequent adjustment to save energy) 5 – User may focus perception on TRV set point at expense of schedule times 6 – User may interpret TRV low setting to indicate no heat is emitted from radiators saved energy 7 – User may overestimate energy saved

(Continued)

TABLE 6.5 (Continued)

How the System Image of the Home-Heating System Can Effect User Mental Models That Underpin Norman's (1986) Seven Stages of Action

Level	Characteristics of Existing UI	Impact of Existing UI on UMMs	Effect on Goal 1: Maintain Routine Comfort Needs (Numbering Relates to Norman's (1986) Stage of Activity)	Effect on Goal 2: Ad hoc Heating (Numbering Relates to Norman's (1986) Stage of Activity)	Effect on Goal 3: Avoid Wasting Heat (Numbering Relates to Norman's (1986) Stage of Activity)
Device level – TRV cont.	Slow response not communicated	• Slow response missing from UMM	3 – User may specify frequent adjustments for TRV to fulfil comfort needs 4 – User may execute frequent adjustments of TRVs 5 – Users may perceive TRV settings as indicators of volume of heat output 6 – User may interpret TRV high setting to indicate heat is emitted at higher temp/rate from radiators	n/a	2 – User may form intention to avoid heating rooms temporarily unused 3 – User may specify frequent adjustments of TRVs to save energy 4 – User may execute frequent adjustments of TRVs 5 – Users may perceive TRV settings as indicators of volume of heat output 6 – User may interpret TRV low setting to indicate no heat is emitted from radiators saved energy 7 – User may overestimate energy saved
	Mapping between setting and room temperature values unclear	• Function of device considered 'valve'	5 – Users may perceive TRV settings as indicators of volume of heat output	n/a	5 – Users may perceive TRV settings as indicators of volume of heat output

(Continued)

TABLE 6.5 (*Continued*)

How the System Image of the Home-Heating System Can Effect User Mental Models That Underpin Norman's (1986) Seven Stages of Action

Level	Characteristics of Existing UI	Impact of Existing UI on UMMs	Effect on Goal 1: Maintain Routine Comfort Needs (Numbering Relates to Norman's (1986) Stage of Activity)	Effect on Goal 2: Ad hoc Heating (Numbering Relates to Norman's (1986) Stage of Activity)	Effect on Goal 3: Avoid Wasting Heat (Numbering Relates to Norman's (1986) Stage of Activity)
Device level – TRV cont.			6 – User may interpret TRV high setting to indicate heat is emitted at higher temp/rate from radiators		6 – User may interpret TRV low setting to indicate no heat is emitted from radiators saved energy
	Distribution across house/flow level access	Meta-data associated with UMM components include effort for access/adjustment	3 – TRV omitted from specification 4 – TRV not adjusted	n/a	3 – TRV omitted from specification 4 – TRV not adjusted

values achieved may differ considerably from the temperature display, so 'comfortable' temperatures may not be possible in all parts of the house at the same time. For the user to be able to fully understand these issues at the time of execution, an understanding of thermodynamics of their home, over time, would also be necessary.

These findings have important implications, as current technology would be capable of addressing these issues. Efforts to address design issues at higher levels would be ineffective, or at best less effective if these underlying issues are not taken on board. This may hinder well-intentioned efforts from achieving the goal of reducing domestic energy consumption, or enabling user goals to be met. For example, improving the usability of the programmer will have little effect if the householder does not include this element in their mental model, and will still waste residual energy if they do not understand the relationship between the time periods set, heat output and comfort goals. Similarly, improving access to TRV controls may mislead users into thinking that these devices provide fast-responding custom heating.

The findings related to home heating such as difficulty in operating controls (e.g. programmer), ambiguous labelling of device controls (e.g. boost button), misleading control design (e.g. thermostat and TRV knob) and the dispersed, unconnected arrangement of related controls were all considered to contribute to issues with both the mental model adopted by users, and their subsequent interaction with the system to achieve their goals.

Norman's (1986) gulf of evaluation and execution to examine home-heating systems at the mental model level has provided insights that could help evaluate recent home-heating system interface applications, as well as inform the design of a digital heating system interface that effectively puts householders in control of their heating goals, whilst reducing wasted energy.

Chapter 7 looks at how a 'control panel' style home-heating interface could be designed that takes into account some of these recommendations, including promotion of the range of controls available, making explicit at the system level their interdependencies, and reducing ambiguities of function.

7 Using Interface Design to Promote a Compatible User Mental Model of Home Heating and Pilot of Experiment to Test the Resulting Design

7.1 INTRODUCTION

This chapter takes theory into practice. It takes the design specification derived in Chapter 6 and applies it to inform the redesign of the home-heating interface. In doing so, this chapter allows the reader to explore the relationship between design and mental models described by hypothesis 3 in the introduction of this book. The pragmatic nature of this chapter embodies the overriding theme (and hypothesis 4) of the book; to use the concept of mental models in design, to elicit behaviour change that results in increased achievement of home-heating goals. The effectiveness of the design changes described in this chapter is tested in an empirical study with a home-heating simulation, and the results reported in Chapter 8.

Domestic consumers contribute to over 25% of total UK carbon emissions (The UK Low Carbon Transition Plan 2009). Large variations in domestic consumption are due to behaviour (Lutzenhiser and Bender 2008). Assumptions about the cause of excessive consumption may include ideas of 'autonomy' ('as long as I can pay for it I should be allowed to use as much as I want'), 'impulsiveness' ('I override the heating as soon as I feel cold'), 'apathy' ('I don't think anything I do makes a difference'), 'lack of confidence' ('It's too difficult to adjust the controls to conserve energy') or 'confusion' ('I don't know what changes I should make, or the changes I make don't seem to have the intended results'). The first three assumptions relate to attitudes to consumption, but the literature struggles to show a strong causal link (e.g. Alwitt and Pitts 1996, Sauer et al. 2009). The latter two assumptions relate to difficulties house-holders have when interacting with energy-consuming devices, and is the focus of an ever-growing body of research (e.g. Kempton 1986, Sauer et al. 2004, 2007, 2012, Vastamäki et al. 2005, Chetty et al. 2008, Lilley 2009, Pierce et al. 2010, Peffer et al. 2011, 2013, Glad 2012, Revell and Stanton 2014).

Domestic space heating accounts for 58% of the domestic consumption in the United Kingdom (Department of Energy and Climate Change 2011) and the central heating system is considered the most complex system in the home (Sauer et al. 2009). Improving householders' confidence and reducing their confusion when operating heating devices would provide the opportunity to reduce wasted consumption. It may also go further and enable users with pro-environmental intentions to demonstrate a significant reduction in consumption, by overcoming the behaviour difficulties often experienced with energy-consuming devices (Kaiser et al. 1999). As Lilley (2009) points out, responsible operation does not necessarily reduce consumption. To reduce consumption by promoting energy-saving goals, it follows that systems would need to enable users to fulfil goals. A system that increased the ability of users to fulfil home-heating goals is therefore desirable.

Lutzenhiser (1993) argues that technology-based efficiency improvements are amplified or dampened by human behaviour. Yet, Glad (2012) found when technology is introduced into homes, it does not always meet the user requirements (in terms of the type of technology and the interface). Pierce et al. (2010) argue that energy-consuming behaviour is unconscious, habitual and irrational, and users ignore visible options. However, Crossman and Cooke (1974) showed slow response systems (such as home heating and cooling) are difficult to learn and control effectively. Sauer et al. (2009)also argue that some users do not have adequate strategies available to manage a system more effectively even if aware of undue consumption. This was borne out in Chapter 3, where it was found that lack of awareness of device controls was responsible for inappropriate strategies with home heating. In addition, Peffer et al. (2011) make the point that to save energy with home heating, the householder must actively select appropriate set point times and expend time and effort in programming devices. They highlight the difficulty multiple users have in reaching agreement on a programmed temperature.

Thinking at the device level, Glad (2012) found thermostats were not used as intended, negatively affecting performance and user satisfaction, making a prudent observation that smart technology is not that smart unless the user can effectively use it. Brown and Cole (2009) support the assertions of Glad (2012) and cite responsiveness, lack of immediate feedback (or lack of relevant feedback) was responsible for poor comfort levels in green buildings. Peffer et al. (2011) found nearly half of users do not use programming features in home-heating devices, and when they do, only half are programmed to make adjustments to correspond with night time or unoccupied times, limiting the energy-saving benefits. They place the cause of misunderstandings of terminology (e.g. set point) and programming difficulties firmly as a result of poor design by designers and engineers. Additional support in the form of operating manuals for programmable thermostats Peffer et al. (2011) found to be lengthy and obtuse, preventing users from easily learning how to operate their controls. This explains Sauer et al.'s (2004) finding that home-heating instruction manuals are often not read by users. Vastamäki et al. (2005) state that most existing temperature controllers do not provide initial feedback in a way that is understandable to the user. In addition, temperature change feedback is delayed, resulting in trial-and-error behaviour, reducing comfort, wasting energy and diminishing user's motivation to operate the control. They describe the temperature control as a seemingly simple everyday device that is difficult to use because 'everyday thinking' leads to the wrong conclusions about its way of working. This echoes the views of

Kempton (1986), who found householders frequently developed an inappropriate 'mental model' of the thermostat, resulting in less appropriate behaviour with that device. It is hoped this chapter has painted a clear picture of the problems household-ers have when operating energy-consuming technology in the home, and highlighted home-heating devices as presenting substantial difficulties. Whilst this summary may appear bleak, the authors believe it clearly directs action to effect change.

Pierce et al. (2010) claim behaviour is strongly influenced by the interface with energy-consuming technology. Lilley (2009) proposes that to sustain change, behav-iour steering should be the strategy adopted by designers. In Chapter 4, Revell and Stanton (2014) argue that designers should focus on misunderstandings and omissions in the user mental models (UMMs) of home heating to encourage users to hold more integrated functional models, and that design strategies could benefit from application of mental model descriptions. Papakostopoulos and Marmaras (2012) proposed that display units should be simplified to reflect the simplified view from expert users. Branaghan et al. (2011) found that users were more effective at performing tasks using a redesigned interface based on expert knowledge, rather than a familiar inter-face. Chapter 6 collected data from a home-heating expert to capture a simplified UMM of the home-heating system and identified how differences between this mental model and the UMMs of novice users led to less appropriate behaviour strategies with heating controls. This follows from Norman (2002), who argues that devices should ensure there is sufficient 'knowledge in the world' to promote appropriate behaviour.

Moray (1990) showed operators accessed mental models when controlling pro-cesses in complex systems. Sauer et al. (2009) considers home heating to be the most complex system in the domestic domain. A mental models approach to home-heating system design therefore warrants exploration. Sauer et al. (2009) when considering instructional displays, emphasized that a poor mental model of system functioning would prevent an operator from knowing how to interpret the information available. Shipworth et al. (2010) echo this sentiment by stating that using controls without understanding how to use them is counter-productive. They propose that new controls should be developed that appeal to householders, are more intuitive to use and make it easy to reduce consumption. Peffer et al. (2011) propose that user misconceptions that encourage incorrect usage cannot be easily overcome by better interfaces. Improved usability is essential but insufficient to encourage correct use, and advocates that a mental models approach to interface design that encourages a compatible UMM at the system and device levels could go some way in promoting appropriate conceptions.

Norman (2002) proposes that differences in user–machine interaction are due to the gulf of evaluation and execution. He argues that interfaces should be designed to promote in the user, a compatible mental model to the design model of the system. To bridge the gulf of execution and evaluation, Norman (2002) advises to 'make things visible' so people know what is possible and how actions should be done, and can also tell the effects of their actions. Chapter 6 applied Norman's seven stages of activity to the operation of a typical home-heating system to understand how devia-tions in householders' UMMs from an 'expert' UMM of the system could result in less appropriate behaviour strategies. The misunderstandings and omissions from these householders' models were tabulated to form a design specification to focus improvements to the home-heating interface with the view of including features that promote a compatible UMM (Revell and Stanton 2016). Manktelow and Jones (1987) reiterate Norman's (1993) argument that good design should lead to a 'single, coherent, and plausible mental model'. They provided 24 recommendations for the design of user–system interfaces by reviewing research focusing on the ways that mental models are invoked and used, including use of language, analogies, salient task features and experience of the user.

The authors have approached the redesign of home-heating controls with the form of a 'control panel' to promote a compatible user mental model (CUMM) of a typical home-heating system. It highlights where the redesign follows generic recommendations from Norman (2002) and Manktelow and Jones (1987) to promote appropriate mental models of home heating in users with specific attention to design specifications described in Chapter 6. The resulting design was subsequently used in a study (described in Chapter 8) to compare how interface design affects the mental models constructed by home-heating users following from Norman (1986), and to compare the impact of mental models of home heating on energy-consuming behaviour as proposed by Kempton (1986). The adaptions made following a pilot run in preparation for this study will therefore also be described, and screen shots of the resulting interface provided.

7.2   CONCEPT DEVELOPMENT

This section will summarize for each key device (thermostat, programmer, boost and thermostat radiator valves (TRVs)) and for the system as a whole, the design specification from Chapter 6. The interface of standard controls, from which these recommendations were derived, will be illustrated and the redesign will be described and illustrated with reference to recommendations for evoking and triggering mental models by Norman (2002) and Manktelow and Jones (1987). The changes made following a pilot for an experiment designed to test the success of the redesign will also be presented.

7.2.1   DESIGN OF KEY DEVICES

7.2.1.1   Thermostat

The design specification from Chapter 6 identified that typical thermostat devices did not effectively communicate the 'feedback' function of the device, nor the temperature sample point. This led to inappropriate functional models of the thermostat and misunderstandings about the scope of temperature control (Revell and Stanton 2014, Chapter 6). Figure 7.1 shows a typical thermostat (based on the Hortsmann – HRT4-ZW thermostat), but showing two temperature values rather than different modes. The control knob is analogous to those found on a 'gas hob'. According to Manktelow and Jones (1987), users' familiarity with this style of control can trigger a 'schema' that the device works in the same way, resulting in a 'valve' mental model of function. Kempton (1986) estimated up to 50% of users in the United States, at the time of his research, erroneously held a 'valve' mental model of the thermostat, resulting in less appropriate operation. Figure 7.2 shows the redesign of

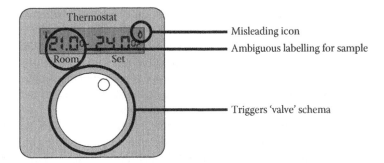

Function: A 'feedback' device that samples the temp. of a single location and sends a message to the boiler to activate/deactivate to maintain the desired temp. set by the user. Boiler activation is conditional on the settings of other integrated devices.

FIGURE 7.1 'Realistic' style thermostat interface with a 'flame' icon indicating boiler operation and ambiguous label 'room' to identify where temperature samples are fed back to the device.

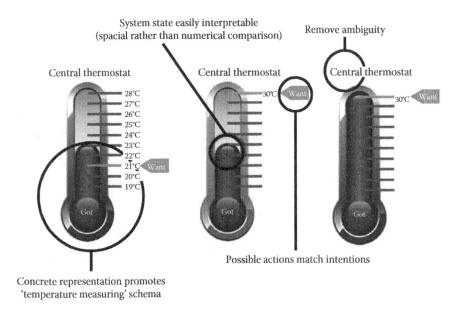

FIGURE 7.2 Redesign of thermostat interface to promote appropriate device model to users.

the thermostat to encourage the idea that existing temperature values fed back from a central temperature sample point are key criteria for device function. Here the design capitalizes on householders' likely familiarity with the form of a thermometer to trigger a schema relating to temperature measurement. Manktelow and Jones (1987) emphasized that ambiguous language should be avoided for the development of coherent mental models. The redesign of the thermostat is labelled a 'central' thermostat, and reference to 'room' is removed to avoid confusion with individual room thermostats. Norman (2002) recommends that users should know what actions are possible and actions should match intentions. The redesign in Figure 7.2 changes the interaction from a rotating knob to a selection of the desired temperature using a 'Want' button. Norman (2002) recommends that the system state is readily perceivable and interpretable and is a match to user's intentions.

The redesign in Figure 7.2 indicates the current temperature measured by the central thermostat by increasing or decreasing the 'mercury' line within the thermostat analogy. In addition, the more meaningful label 'got' (as opposed to 'room') is used. By placing the 'got' and 'want' values on a single temperature scale, a concrete and immediate comparison of these values is apparent to the user. The functioning of the thermostat relies on this comparison of values (if the 'want' value is higher than the 'got' value, the thermostat 'calls for heat'). Norman (2002) recommends that the effects of their actions should be visible to the user. In the redesign in Figure 7.2, the outline of the thermostat changes to green when calling for heat. The authors deliberately avoided using a flame icon (as with the typical design in Figure 7.1), to prevent users being misled that an 'active' thermostat alone equates to boiler activation. Norman (2002) recommends that the outcome of users' actions should be obvious. As boiler activation relies on additional devices, how the redesign follows this recommendation is discussed in the 'system' section.

7.2.1.2 Programmer

The design recommendations from Chapter 6 for the programmer identified that the usability issues of typical programmer devices create 'meta-knowledge' (Bainbridge 1992) in UMMs that dissuades the inclusion of programmers in the behaviour strategies formed by users, as they require too much effort for operation. In addition, the conditional link to the thermostat for operation is not clearly communicated, which can lead users to assume that scheduled times equate to 'boiler on' times (Revell and Stanton 2014). Finally, the difference between scheduled end times and continued comfort levels due to residual heat is not highlighted, preventing thermodynamic lag times forming part of UMMs that could promote shorter programmer schedule times. Figure 7.3 shows a typical programmer (based on a Horstmann CentaurPlus 17).

Figure 7.4 shows a redesign of the programmer to overcome lack of use due to usability issues. Norman (2002) suggests that the structure of tasks can be simplified by minimizing the amount of planning necessary, and changing the nature of the task. The redesign in Figure 7.4 changed the nature of the task from navigating through a series of modes and options to input start and end times, to a simple 'point and click' task to select hour slots of operation. Following from Norman's (2002) recommendations, it makes visible the actions that are possible, how the action is to be done, and unlike traditional programmers (e.g. Figure 7.3), it visibly displays the programmed schedule resulting from their actions by changing the time slots to green. To indicate the system state, when a scheduled time slot is 'active', the outline of the programmer is highlighted in green (Figure 7.4), supporting evaluation of system state. Promotion of the conditional link and thermodynamic lag times are discussed at the 'system level' as they relate to other devices and target rooms, respectively.

Unintuitive multi-step task promotes negative 'meta-data'
in users mental model

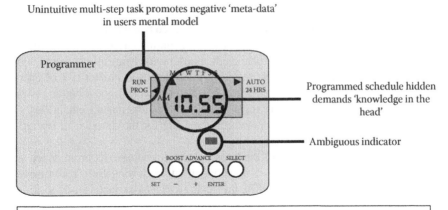

Programmer

Programmed schedule hidden
demands 'knowledge in the
head'

Ambiguous indicator

Function: An automatic 'on/off' switch that sends a message to the boiler to activate/
deactivate according to a pre-programmed schedule. Boiler activation is conditional
on the settings of other integrated devices.

FIGURE 7.3   'Realistic' home-heating programmer interface. Red indicator illuminates during scheduled 'on periods'.

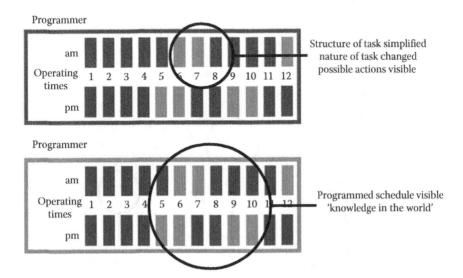

Programmer

am

Operating   1   2   3   4   5   6   7   8   9   10 11 12
times

pm

Structure of task simplified
nature of task changed
possible actions visible

Programmer

am

Operating   1   2   3   4   5   6   7   8   9   10 11 12
times

pm

Programmed schedule visible
'knowledge in the world'

FIGURE 7.4   Redesign of programmer to simplify schedule input, to encourage inclusion in behaviour strategies.

7.2.1.3 Boost

The design recommendations from Chapter 6 identified that the function of the boost button was not communicated clearly, nor was its conditional link to the thermostat for boiler operation. The former explained the boost button being absent from UMMs or avoided in behaviour strategies, resulting in strategies at greater risk of energy waste (Revell and Stanton 2016). Figure 7.5 shows a typical boost button, as part of the features on the programmer (again based on the Horstmann CentaurPlus 17). When pressed, 'BOOST' text is shown on the LED for the duration of operation (1 hour).

Figure 7.6 shows a redesign of the boost feature to increase its prominence as a key control, and communicate its 1-hour operation. Following from recommendations by Manktelow and Jones (1987), the analogy of a clock face is used to trigger

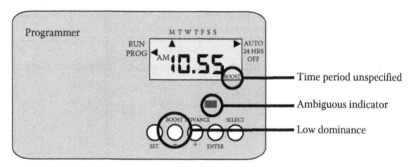

Function: A manual 'on', automatic 'off' switch device that sends a message to the boiler to deactivate after a specific time period (1 hour) has elapsed. Boiler activation is conditional on the settings of other integrated devices.

FIGURE 7.5 'Realistic' boost button, as a feature on a programmer device. Boost text appears in the LED display when active.

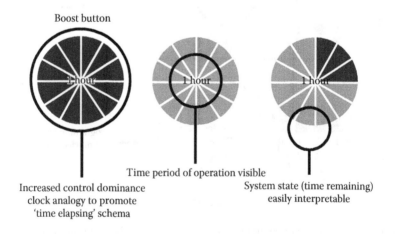

FIGURE 7.6 Redesign of boost button to promote 1-hour operation.

in the user a schema relating to time passing and a label of '1 hour' is present to avoid ambiguity in the time period of activation (Figure 7.6). This also supports Norman's (2002) recommendations to make visible to users what actions are possible (only a 1 hour option). When activated, the time remaining for operation is displayed in green, following Norman's (2002) recommendation to make visible the effect of user's actions to enable easy evaluation of the system state. For promotion of the conditional link represented at the 'system level', see Figure 7.10.

7.2.1.4 TRV

The design recommendations from Chapter 6 highlighted problems in the design of traditional TRVs in the communication of the feedback function, the mapping between set point and room temperature values and their slow responding nature. This was found to result in an inappropriate device model and action specification (Revell and Stanton 2016) for the TRV. In addition, the distributed ankle-height positioning was thought to encourage meta-knowledge (Bainbridge 1992) in UMMs that dissuade inclusion in users' behaviour strategies due to the physical effort involved in adjustment (even if the correct device model was held). Figure 7.7 shows a typical TRV control (based on a 15 mm Ravenheat TRV). The set point scale shows markings from 1 to 5 without mappings to temperature values. This control suffers from the same difficulties as the central thermostat, in that a rotating knob is likely to evoke schemas based on users' experience with alternate devices that function like a value (e.g. Gas hob, basin tap). According to Manktelow and Jones (1987), the consequence of this could be the development of an inappropriate mental model (e.g. valve), as was found in Revell and Stanton (2016).

Figure 7.8 shows a redesign of the TRV device, following recommendations from Norman (2002), the mapping between set points and approximate room temperatures are made explicit, to better enable users to match actions to intentions. Echoing the

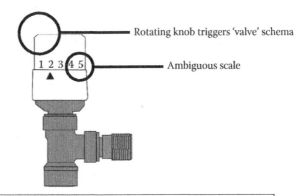

Function: A slow responding 'feedback' device that limits heat by allowing/blocking hot water to flow into the connected radiator. It samples the temp. at its location. if higher than the associated temperature range corresponding to the set point, it blocks water. If within, or lower, it allows water to flow.

FIGURE 7.7   'Realistic' style of TRV control, with ambiguous 5-point scale.

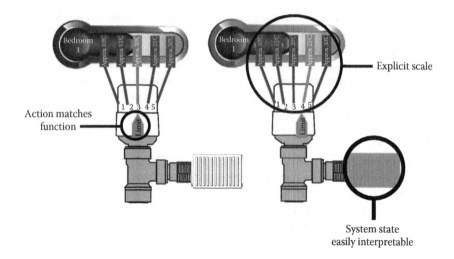

Explicit scale

Action matches
function

System state
easily interpretable

FIGURE 7.8   Redesign of TRV controls to promote heat-limiting feedback device model.

redesign of the central thermostat and following from Manktelow and Jones (1987), the image of a thermometer is used to evoke the concept of temperature measurement, emphasizing data fed back to the device is essential for operation. To distinguish its operation from the way the thermostat functions, the set point button is labelled 'Limit' to communicate that the device limits rather than achieves room temperatures. To promote that TRV controls act on individual rooms, the target room is labelled in the mercury line to indicate the temperature value in that room. Similar to the central thermostat, the comparison between the room temperature in the target room and the temperature at which heat output from the radiator will be limited is visually emphasized (Figure 7.8). Recommendations from Norman (2002) to make visible the possible actions are clear from the choice of five settings. The effects of users' actions is displayed visibly by highlighting the chosen setting, and where this has not been reached by the 'mercury line' for the target room, the outline of the thermometer as well as the radiator image are highlighted in green, providing a visible and readily interpretable state of the device. The slow responsiveness of the TRV was emphasized by operation instructions on the interface and in the user manual. Negative meta-knowledge was expected to be overcome by a control panel interface where all TRV controls can be accessed without extensive physical effort.

7.2.2   DESIGN OF SYSTEM VIEW

The design recommendations from Chapter 6 highlighted how a typical system view of home heating has characteristics that impeded development of an appropriate UMM by householders. Presenting a range of different controls of varying prominence distributed around the house risks some controls being overlooked (e.g. Revell and Stanton 2014), the hierarchy of controls being misunderstood (Revell and Stanton 2016) and meta-data for effort involved in controlling or checking device

'Tunnel vision' system state (e.g. comfort levels/radiators) can only be evaluated one room at a time

Device control/device feedback only possible individually when co-located

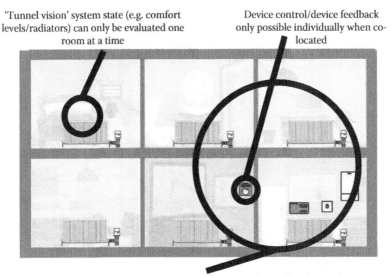

Devices distributed across home with varying prominence/hierarchy of controls
Lack of visible connections between devices preventing interdependencies being understood

FIGURE 7.9 Distribution of typical home-heating controls across the home (but users can typically only see one room at a time).

status preventing appropriate actions with, and evaluation of, the heating system (Revell and Stanton 2016). The lack of household thermodynamic data in typical home-heating systems hinders this concept becoming part of UMMs affecting expectations of performance (Revell and Stanton 2016). A lack of clear visible connections between devices prevents interdependency of devices being emphasized and cause-and-effect rules being developed (Revell and Stanton 2016). A typical distribution of controls is shown in Figure 7.9. Access to these controls, as well as feedback of their status, and the status of comfort levels in each room are experienced by the user in what is best described as 'tunnel vision'. The householder can only be present in one room at a time, meaning they can only access devices for control and feedback when present in the same room. Similarly, they can only evaluate the progress with comfort levels for the room in which they are situated, limiting an overview of variations in comfort levels across the house. This potentially hinders the development in UMMs of a conceptual representation of thermodynamic variations.

Norman (2002) emphasizes that to bridge the gulf of execution and evaluation, the outcomes of an action should be obvious. Figures 7.10 and 7.11 show the redesign of the home-heating interface to display the link between distributed physical control devices around the home (Figure 7.10) and to emphasize the 'conditional rule' that integrates set point choices for key controls (thermostat, programmer and boost) at the point of action (Figure 7.11) when controlling boiler operation.

Following recommendations by Manktelow and Jones (1987), an analogy of a 'switches in a circuit' is used to trigger a schema that guides the development of a

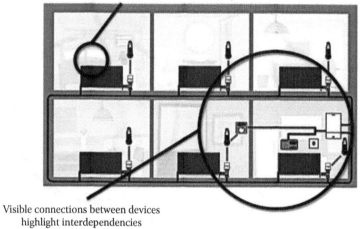

FIGURE 7.10 Redesign of interface to display distributed control devices with cause-and-effect links.

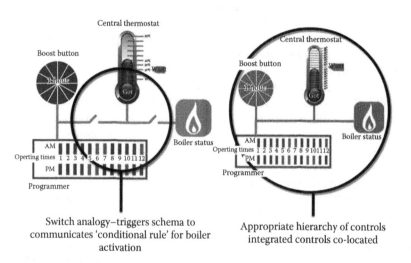

FIGURE 7.11 Control panel for key controls emphasizing link between set point choice with key controls and boiler status.

functional UMM (Figure 7.11). Boiler activation is made visible (highlighted green) when the programmer or boost is active (closing left switch) *and* the thermostat is active (closing right switch). If one or both of these conditions are not met, one or both switches will be open and the boiler icon will remain inactive (grey).

To aid discoverability of controls – redesigning the interface as a 'control panel' enabled access to all controls from a single point. Manktelow and Jones (1987) stress that if appropriate dominance is not emphasized, correct assumptions about dominance in development of UMMs is unlikely. The hierarchy of controls was emphasized by placing the key controls visible on the main panel. The only exception to this was the TRV controls. As slow-responding long-term controls, they are not suitable for frequent adjustments. Norman (2002) advocates deliberately making things difficult if there are undesirable consequences for operation. Access to the TRV controls were therefore placed with the advanced controls via a button on the main panel. This button categorized lesser used controls by their function with the option to place further text instructions at the point of action (see final design of the main panel shown in Figure 7.13). In the redesign of the interface, 'tunnel vision' is removed by showing the comfort levels of all rooms in a single view. This follows from Norman's (2002) recommendation to ensure the system state is visible and readily interpretable. The intention of the design is that variations in comfort levels across the house will promote an understanding of lag times and thermodynamic variations in user's mental models of home heating. The room's overview in Figure 7.10 also displays the location of physical controls within the home and makes the invisible visible (Norman 2002), by showing the links between devices. Thermometer icons linked to the TRVs and central thermostat are also indicated to reinforce the feedback function of these devices relies on temperature sensing. To make visible the outcomes of users' actions with TRVs following recommendations from Norman (2002), the radiators are highlighted in green (Figure 7.10) when water can flow, but not when the TRV acts to limit heat. In the final iteration of the simulation design, a text label 'water flow allowed' or 'water flow blocked' is also added to the simulation design to reinforce the TRV function (Figure 7.13).

7.2.3 CREATING A SIMULATION

A simulation was created to enable the redesigned controls and interface to be compared in operation to a more realistic interface (based on the typical controls and interface described). The simulation presented two versions of a home-heating interface, with controls providing the same function 'behind the scenes'. The realistic interface replicated the 'tunnel vision' experience by presenting a single room to users at a time, meaning feedback of the devices available, and the comfort levels achieved were only possible for the visible room (see Figure 7.12). To reflect users' need to deliberately approach a device before operation, the representation of typical devices in the typical room could only be accessed when the user clicked the correct room (e.g. 'entered the room') and the specific device (e.g. approached the device). A larger version of the device was then presented that allowed adjustment with the mouse. This is in contrast to the redesigned simulation that provides visual feedback of all rooms and access to key controls by default (Figure 7.13).

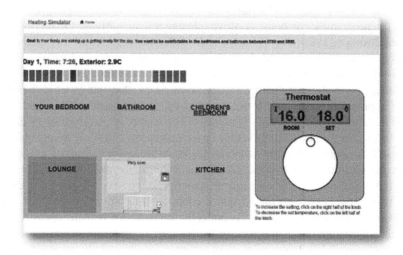

FIGURE 7.12   Interface created to represent a typical home-heating interaction.

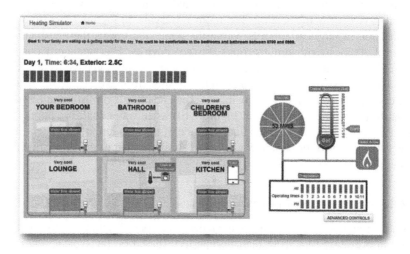

FIGURE 7.13   Interface designed to promote compatible user mental model of home-heating system.

To avoid differences in opinion by users as to what temperature values constituted 'comfortable', feedback was provided with descriptions ranging from very cold, cold, too cool, comfortable, too warm, hot and very hot. The room temperature range for each of these descriptions was adjusted to ensure the user could not achieve comfortable temperatures too easily. This was necessary as a link between mental models and behaviour under investigation, so a motivation for adjustment was needed to capture any differences. A simple thermodynamic model was included whereby the rooms adjacent to outside walls lost heat at a higher rate than central rooms, and included changes to heat loss rates based on variations in external temperatures throughout the day. Some elements of the interfaces shown in Figures 7.12 and 7.13 do not relate to promotion of compatible UMMs, but are purely a component of the experimental setting. These include the text presentation of goals, and the 'ticker' below this that shows progress through the experiment.

7.2.4 PILOT

Initial informal pilots were undertaken to identify and mitigate initial usability issues during interface development. Following this, a formal pilot for each condition was undertaken. This ran through the whole experimental procedure, and provided feedback about the experience from a participant's point of view. The pilot candidates were both PhD students in engineering and the environment in their mid-20s. The issues and mitigation strategies that emerged from these formal pilots are summarized in Table 7.1.

The resulting interface design for both the realistic and redesigned versions are shown in Figures 7.12 and 7.13. The intended redesign for the boost button (relabelled 'Override') as shown in Figure 7.6 was not possible to implement in the time allocated. The resulting design showed all segments highlighted as green when active, with the remaining time displayed on the text label (Figure 7.13).

7.3 DISCUSSION

Home heating control, like other devices in the domestic domain, suffers from the difficulty that users are given no formal training on operation and instruction manuals may not be consulted, are too difficult to understand or missing (Kempton 1986, Sauer et al. 2009). Whilst the impetus for this research is to help reduce energy consumption due to space heating, the focus in this chapter is considered to be the first step towards this goal. The aim of the research is to change users' mental models through the way in which heating controls are presented to householders. It is hypothesized that the design and controls will promote more compatible models than traditional interfaces.

The design decisions made were directed from the design specifications to improve mental models of home heating, provided in Chapter 6, and design recommendations to evoke mental models in general, by Norman (2002) and Manktelow and Jones (1987). Some aspects of the designs intended could not be fulfilled (e.g. Boost (Override) redesign). Other design elements that made sense 'on the drawing board' did not work in practice (e.g. providing a link between key control devices on the house display, as well as a feedback link from thermometer icons to TRVs), highlighting the benefit of piloting design concepts. Other design considerations were made to ensure the use of a simulation for an experiment to test differences in behaviour resulting from altered UMMs provided sufficient data. This included not only potentially narrower than necessary criteria for room temperature values labelled as 'comfortable', but also a deliberate decision to enable unhindered access to the thermostat and programmer controls. These specific controls are considered

TABLE 7.1

Summary of Issues and Changes to Experimental Procedure, Interfaces and Setting, Following Pilot Run

Setting

Issues	Mitigation
1. Awkward orientation to screen	1. Position monitor face on to participant.

Procedure

Issues	Mitigation
1. Key elements of the simulation (unaware/ unclear about the purpose)	1. During 'play mode', explain layout, accelerated time, changing goals, temperatures in rooms changing is also due to thermodynamic model, and purpose of 'ticker' is to show progress through simulation. Make goal change more noticeable with screen flash/colours and make key goal information bold.
2. Interaction with simulation (prevalence and functioning of goals)	
3. Manual use (unsure of expectation)	
4. Goals (concern over expectations, disbelief about achievability)	
5. Stressful experience	2. During 'play mode', make clear there are controls in each room and that all controls represented can be interacted with and are working.
	3. Emphasize that manuals are there to explain control operation and can be used as much or as little as needed in the 'play' and 'main' experiment.
	4. During 'play mode', talk through goal example. Make clear prior to experiment condition that (a) goals are achievable but can be difficult, (b) all goals should be attempted, (c) if goal is already achieved or existing settings are appropriate, no action is necessary.
	5. Prior to experiment condition and during debrief, reassure participant that goals may be difficult to achieve but that the aim of the experiment is to see the controls used as settings chosen, rather than if the goals were achieved.

Realistic interface

Issues	Mitigation
1. Usability of interaction for adjusting thermostat, boiler water control, TRV	1. Put text instructions next to controls and warn participant that some controls are difficult to operate with the mouse.
2. Mismatch between room labels and goal reference	2. Change room labels to match goals (e.g. 'bedroom one' = 'your bedroom'.
3. Misleading feedback of room comfort levels	3. Change labels to 'too warm' and 'too cool' rather than 'warm' and 'cool', so participant knows that goal has not been achieved.

(Continued)

TABLE 7.1 (*Continued*)

Summary of Issues and Changes to Experimental Procedure, Interfaces and Setting, Following Pilot Run

Design interface	
Issues	Mitigation
1. House image too cluttered – difficult to know what to focus on 2. Boost label misleading 3. TRV purpose not clear 4. Not clear you can interact with thermostat	1. Remove from house image: thermometers and black line link to the radiator controls, lines linking programmer/thermostat to boiler and programmer and power switch icons. 2. Rename boost as 'override'. 3. Label TRV control so function explicit. 4. Emphasize thermostat buttons.
Achievability of task	
Issues	Mitigation
1. Goals too easy/hard to achieve 2. Boiler 5°C working – makes accidental messing around have too high an effect	1. Experiment with parameters for 'comfortable' temperature setting. 2. Make boiler 5°C setting a 'dummy' so pressing it can be recorded as an action, but it does not have an effect on the simulation.

long-term controls requiring initial and careful setup (Revell and Stanton 2016). For application in a genuine domestic setting, it is recommended that different modes of access to these controls, depending on whether the user is using trial and error to determine appropriate set points to meet their needs (full access), or if long-term set points had already been established, and the user intends only to make adjustments for atypical goals (restricted access).

The body of literature referred to in the introduction, focusing on problems with energy-consuming devices in the home, also offered up a number of recommendations for improving users' interactions with technology. It is put forward that a mental models approach to design at both the system and device levels is likely to address a number of these. For example: real-time feedback, in this design, relating to comfort levels and boiler on periods (Sauer et al. 2004, Chetty et al. 2008); clear display of device status (Sauer et al. 2007); improved discoverability and intuitive design of controls (Glad 2012); understandable temperature scales, in this case for the TRV controls (Vastamäki et al. 2005); preference for auto-reset features, supported in this design by providing increased prominence of the boost button (Sauer et al. 2004); peripheral positioning of controls (in this case TRV and advanced controls) to reduce frequency of adjustment (Sauer et al. 2004) and application of information proximal to controls (Sauer et al. 2004). In addition, Pierce et al. (2010) and Peffer et al. (2013) validate the relevance of generic design recommendations by Norman (2002) by advocating designers make use of affordances and constraints, visibility, feedback, natural mappings and consistency when ensuring householders' effectively operating domestic energy-consuming devices. These, together with other recommendations in the literature, can be an asset. However, designers need to ensure that seemingly separate devices that in practice have to be operated with consideration of the impact on each other are 'improved' with consideration of the user's mental models at a system level.

7.4   CONCLUSION

This chapter has highlighted how domestic technology, particularly the central heating system, is problematic for the user. A case was made for redesigning the home-heating interface so that an appropriate user mental model was projected to the user. A redesign of home-heating controls at the system and device levels was undertaken, based on a design specification developed in Chapter 6. Recommendations by Norman (2002) and Manktelow and Jones (1987) for evoking appropriate UMMs were applied to the designs. A brief description of the development of a simulation to test the effectiveness of the redesign, as well as changes resulting from a pilot was provided. Limitations of the design, as well as further enhancements were explored. How the existing design meets recommendations for improving interaction with energy-consuming devices in the home was discussed. It is proposed that the designer consider the redesign of home-heating controls at the system level to evoke functional mental models in the user. These ideas are tested in Chapter 8.

8 Mental Model Interface Design

Putting Users in Control of Their Home-Heating Systems

8.1 INTRODUCTION

The final chapter in this book considers all the hypotheses described in Chapter 1 in an empirical study and draws together threads developed in the preceding chapters. It explores how a mental-model-driven design can influence the model held (hypothesis 3), which in turn influences behaviour patterns (hypothesis 1) to improve the achievement of home-heating goals (hypothesis 2). Linking these hypotheses together with the same data source provides evidence to test overall hypothesis 4 that 'the knowledge of existing mental models of devices can be used in device design to encourage patterns of device use that increase goal achievement'. This chapter performs statistical tests on data collected through application of QuACk developed in Chapter 3, as well as automated data collected from a home-heating simulation. The simulation design is documented in Chapter 7 and is informed by the design insights identified in Chapter 6. These, in turn, were based on evaluation of the omissions and errors in householders' mental models of home heating described in Chapters 4 and 5.

Energy consumption due to home heating is a key contributor to climate change, making up 58% of UK domestic energy consumption (Department of Energy and Climate Change 2011). It is easy to save energy in the home: just don't turn the heating on (Sauer et al. 2009). The real challenge is *using* energy effectively and efficiently, not just saving it. Using energy effectively to meet realistic heating goals is no mean feat (Revell and Stanton 2014, Chapter 4). Doing so in a way that minimizes waste is rife with difficulties when using devices that were not designed with this emphasis (Sauer et al. 2009, Revell and Stanton 2016).

Occupant behaviour is a key variable affecting the amount of energy used in homes (Raaij and Verhallen 1983, Emery and Kippenham 2006, Lutzenhiser and Bender 2008, Guerra-Santin and Itard 2010, Dalla Rosa and Christensen 2011). Kempton (1986) discovered that variations in the way occupants behaved with their home-heating thermostat could be explained by differences in their 'mental model' of its function. Different types of mental models held by occupants encouraged different behaviour strategies for saving energy overnight. Kempton (1986) estimated considerable energy savings could result if specific mental models of thermostat function were promoted to domestic users. Householders' misunderstandings about thermostat function and the workings of the heating system as a whole is still a problem today, however (Brown and Cole 2009, Shipworth et al. 2010, Revell and Stanton 2014, 2015). Revell and Stanton (2014, 2015, 2016) extended the findings of Kempton (1986) to consider functional mental models of all home-heating controls present in the home, as well as their interactions at a system level. In addition to inappropriate mental models of device function, in Chapter 4 it was found that incomplete mental models at a system level explained differences in behaviour strategies that either wasted energy or jeopardized comfort goals.

Mental models are described as internal representations of the physical world (Veldhuyzen and Stassen 1976, Johnson-Laird 1983, Rasmussen 1983) and can act as an internal mechanism to allow users to understand, explain, predict and operate the states of systems (Craik 1943, Gentner and Stevens 1983, Kieras and Bovair 1984, Rouse and Morris 1986, Hanisch et al. 1991). The link between mental models and the operation of states of systems allows these 'internal representations' to help explain human behaviour (Gentner and Stevens 1983, Wickens 1984, Kempton 1986). There are many definitions of mental models and different perspectives from which to consider them (Wilson and Rutherford 1989, Richardson and Ball 2009, Revell and Stanton 2012), so specificity in definition is key (Bainbridge 1992, Revell and Stanton 2012). In this chapter, the concept is best understood in terms of a user mental model (UMM) (Norman 1983) and device model (Kieras and Bovair 1984). That is to say, a mental model held by a user of a specific technology that contains information about the operation and function of that device, and has been accessed and described by an analyst.

Lutzenhiser (1993) argues that human behaviour limits the efficiency of technology introduced to reduce consumption. However, it is often the choice and positioning of technology, as well as usability issues that impede discovery and use by householders (Glad 2012). As the case was made in Chapter 6, poor discoverability of controls is a credible cause of incomplete models (Shipworth et al. 2010, Revell and Stanton 2014) and usability issues of home-heating controls impede effective operation (Brown and Cole 2009, Combe et al. 2011, Shipworth et al. 2010, Peffer et al. 2011, 2013, Glad 2012). These can result in lack of use due to inconvenience (Chapter 5) or fear of complexity (Glad 2012). As background knowledge gained from experience affects the formation of the user's mental model (Johnson-Laird 1983, Moray 1990, Bainbridge 1992), lack of use can further impede appropriate models of heating controls, leading to inappropriate behaviour that further impacts home-heating goals (Kempton 1986).

Norman (1986) proposed that designers could help users operate technological systems more appropriately by designing interfaces that encourage a 'compatible' mental model of the way the system functions. Norman (1986, 2002) proposed that a compatible mental model was necessary to enable users to successfully navigate the 'gulf of evaluation and execution' when interacting with a system. In Chapter 6, the home-heating system was considered from the perspective of Norman's (1986) seven stages of activity, to determine the components necessary in a compatible mental model for typical home-heating goals. Norman (1983, 1986) emphasized that whilst underpinned by their mental model, users' interaction with technology are ultimately driven by their goals. This is supported by Bainbridge (1992) and Moray (1990) who argue that users' mental model of the system constrains the performance with the system, but user goals influence the resulting strategy adopted.

The review so far has touched on system device design (in terms of discoverability and usability of controls), mental models (at device and system levels), user goals (both for comfort and energy saving) and the strategies adopted (in terms of operation of controls). These are all variables that influence the observed user behaviour with home-heating systems (for a simplified view of how these variables interact, see Figure 8.1). The consequences of user behaviour need to be understood in relation to the intended goals. However, as both Kempton (1986) and Sauer et al. (2009) note, goal achievement is also subject to variables acting in the broader system (e.g. building structure, infiltration, insulation, thermodynamics of the house, external temperature), as depicted in Figure 8.1.

In a realistic setting, there are clear barriers to controlling householders' goals and broader system variables, limiting the insights that can be drawn about the cause of differences in occupants' behaviour. Sauer et al. (2009) overcame these issues with a central heating system simulator that enabled control of house structure, external weather conditions, occupancy and regularity of arrivals and departures. Sauer's (2009) study compared the provision of different types of feedback on consumption levels (revealing predictive feedback to be the most effective at reducing consumption). Sauer's (2009) participants were all university students, and were tasked with creating heating profiles for each room within the simulation, by selecting heating

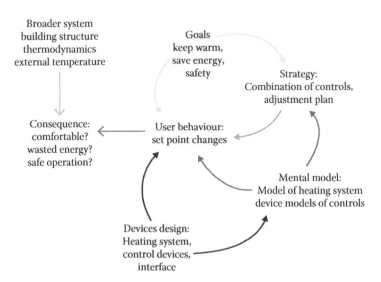

FIGURE 8.1 Different variables that effect home-heating behaviour and its consequences.

periods on a graph-style interface. This differed to the nature of typical home-heating goals, and for greater generalizability of results, Sauer et al. 2009 highlighted that a broader age group and more realistic goals were needed.

Previous work has revealed insights about UMMs of home heating at the individual level from a small, atypical sample selected for minimal experience with home-heating control (Revell and Stanton 2014). Use of a simulation in this study was adopted to allow a larger sample of typical home-heating users in the United Kingdom, enabling greater generalizability of conclusions. In addition, the simulation allowed control of goals, broader system variables and heating system type.

This chapter investigates how differences in interface design affect UMMs of home-heating systems. Using data from a simplified home-heating simulation, it explores how resulting changes in UMMs affect user behaviour with home-heating controls. Finally, differences in goal achievement resulting from different behaviour strategies enabled by different interfaces are compared. Classic studies focused on manipulating users' mental model and comparing their behaviour and performance (e.g. Kieras and Bovair 1984, Hanisch et al. 1991) with a single device. Using an extensive 'training' stage to alter the mental model, change is presumed to have occurred if the expected performance was observed. This study differs in that it relies not on prior training, but on the interface design (plus manuals) to alter the mental model held *at the time of interaction with the system*. This study also measures the success of this manipulation by analysis of user-verified UMM descriptions captured post study, behaviour related to statistically significant differences in UMMs and overall goal achievement. Following from the work of Norman (1986, 2002), the premise of the study predicts that an interface designed to promote a compatible mental model of the heating system, better enables users to achieve their home-heating goals.

8.2 METHOD

The findings over the preceding chapters summarized in the introduction require updates to the original hypotheses set out in Chapter 1. The perspectives by which mental models, behaviour and goal achievement are considered need to be better specified, and deserve definition to the reader. Each of the original hypotheses 1–3 is tackled in this chapter through focus on more granular 'sub-hypotheses' relating to the variables collected. Figure 8.2 maps the sub-hypotheses onto the updated original hypotheses, so it is clear how the results contribute to the original intentions of this research. This diagram also illustrates well the 'cascading' nature of the hypotheses, which informs the experimental design.

Whilst slightly beyond the scope of the main focus of this chapter, the study undertaken provided an opportunity to test an overriding assumption for this research: that the contents of UMM descriptions captured by the Quick Association Check (QuACK) developed in Chapter 3 (as opposed to other methods) are linked to behaviour with home-heating controls. The results of this investigation are also shown in this chapter, to add validity to the novel method deployed. The key focus, however, predicted that participants presented with an interface designed to promote

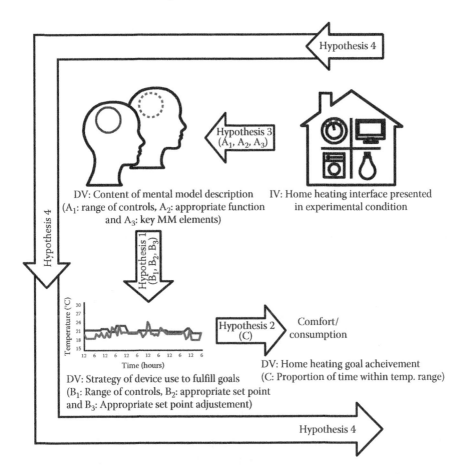

FIGURE 8.2 Diagram to show relationship between theses hypotheses and specific hypotheses tested in the thesis. The linked hypotheses informed the 'cascading' experiment design.

appropriate UMMs would exhibit positive differences in their mental models, behaviour strategies and goal achievement.

8.2.1 Experimental Design

The experiment was a between-subjects design. The independent variable was the version of the interface of a home-heating simulation presented to participants: either 'realistic' or 'design' (see Figure 8.3). Dependent variables related to participants' UMM, their behaviour with controls to achieve goals and the level of goal achievement in line with the hypotheses stated in the following.

Hypothesis 3 states that 'the device design influences mental models of those devices'. To test this hypothesis, dependent variables relating to UMMs included (1) the range of heating controls present in participants' UMM description, (2) the number of appropriate functional models of key controls and (3) the number of key

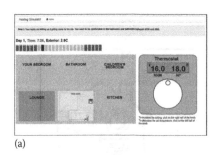 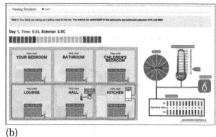

(a)   (b)

FIGURE 8.3   Screen shots of realistic interface (a) and design interface (b).

mental model elements for home-heating operation present in participants' UMM descriptions (see Figure 8.1). Relating to hypothesis 3, this chapter tests the following sub-hypotheses:

1. *Hypothesis* (A_1): Users would have a greater range of heating controls in their mental model description (through enhanced discoverability recommended in Chapters 6 and 7).
2. *Hypothesis* (A_2): Users were more likely to hold an appropriate functional model of key devices (through redesign of controls, recommended by Norman (1986, 2002)).
3. *Hypothesis* (A_3): Users would describe mental models with a greater number of key home-heating system elements (through redesign of the home-heating interface at a system level (see Chapters 3 through 6).

Hypothesis 1 from Chapter 1 stated: 'Users' mental models of devices influence their pattern of device use'. Following the insights gained in the preceding chapters, this hypothesis has been amended to: 'Users' mental models of devices influence their *strategies* of device use'. Dependent variables relating to users' behaviour with the simulation included (1) the range of controls used in participants' strategies and (2) the number and value of set point adjustments time (see Figure 8.1). The following sub-hypotheses are tested to inform Hypothesis 1:

- *Hypothesis* (B_1): Where significant differences in the presence of specific devices in UMMs were found, associated differences would be found in the inclusion of those devices in behaviour strategies.
- *Hypothesis* (B_2): Where significant differences in the appropriateness of functional models at the device and system level were found, associated differences would be found in the adoption of set point values and frequency of adjustment for those devices (following from Kempton (1986) and Norman (1986, 2002).

Hypothesis 2 from Chapter 1 stated: 'Patterns of device use influence the amount of energy consumed over time'. This hypothesis has been amended to '*Strategies* of device use influence the amount of *goal achievement*'. This was done for two reasons. First, the goals used in the simulation were derived from interview data for realism, and focused more on comfort than consumption. Second, the simulation was not sufficiently sophisticated in its modelling to provide a meaningful comparison of boiler 'on' times. The dependent variable used to represent goal achievement was that the proportion of time target room temperatures were within the target temperature range (see Figure 8.1). This chapter contributes evidence to hypothesis 2 with the following sub-hypothesis:

- *Hypothesis* (*C*): Duration of goal achievement would be greater for participant in the 'design' condition, than for participants presented with the 'realistic' interface (following from Norman 1986, 2002, Moray 1990, Bainbridge 1992).

Figure 8.2 illustrates how each hypothesis is linked to the other, resulting in a 'cascading' experiment design, where goal achievement variables are dependent upon the outcomes of the behaviour variables, which are in turn dependent upon the mental model variables. Where an unexpected outcome from the mental model or behaviour variables occurred, the expectations for behaviour and goal achievement outcomes, respectively, were therefore amended.

8.2.2 PARTICIPANTS

Forty participants took part in this experiment, 20 per condition; 10 males and 10 females were in each condition from ages ranging between 23 and 70 years (Mean = 38). Pairs in each condition were matched by gender, age category and the number of years' experience with central heating (±2 years). Experience with gas central heating (with radiators) ranged from 4 to 40 years, with a median of 12 years. Participants were all native English speakers and were recruited from staff, students and residents local to the University of Southampton. Participants were recruited through posters on university notice boards and website.

8.2.3 APPARATUS AND MATERIALS

Development of the simulation allowed automatic data collection of energy use and behaviour strategies, and enabled two different interface views to be presented to participants. The simulation versions were presented on a Samsung LE40M67BD 40″ TV monitor attached to a DELL Latitude E6400 laptop connected to the Internet page hosting the simulation and controlled with a mouse. The 'realistic' version reflected the design of a typical gas central heating system, using the setup described in Chapter 3 and Revell and Stanton (2014). The 'design' interface was constructed to promote a compatible mental model of the home-heating system, following recommendations from Chapter 6. The duration of the simulation activity, goals presented and duration of goals were matched for both conditions. The simulation automatically collected data relating to (1) set point adjustments (2) 'boiler on' periods and (3) 'device control mouse clicks'. A 'play mode' was available for each version of the simulation that did not collect data. User manuals specific to each simulation version were provided. A consent form, participant information sheet and participant instructions were provided along with a pen for the subject. Parts 1 and 3 of the QuACk interview script (Appendix A) were used to interview participants to collect background information and elicit the users' mental model. The former was recorded using a pen, and the latter was recorded on A3 plain paper and square post-it notes, using a marker pen. The interview was recorded on an Olympus VN-2100PC audio recorder.

8.2.4 PROCEDURE

Subjects undertook the study individually, sitting at a desk in front of the monitor. After providing health and safety information, the subject was asked to fill in the consent form, read instructions for participation and fill in the demographic information section on this sheet using the pen provided. The experimenter then checked understanding, verbally reiterated the key points of the experiment, told the subject which condition they were allocated to and collected the completed consent forms.

Before starting data collection, the participant was provided with the user manuals for their experimental condition and exposed to a 5 minute practice session, using their simulation version in 'play' mode. At the start of the practice session, the experimenter pointed out key elements of the interface with reference to a script appropriate to the experimental condition. Those in the 'realistic' condition were asked to imagine they were operating the home-heating controls as if they were in their own home setting. For the 'design' condition, they were asked to imagine they had been provided with a digital interface to control the existing heating controls in their home setting. The experimenter then remained at a desk behind and to the left of the subject to allow the subject to practise independently. During the practice session only, the experimenter responded to any questions by the participant about the procedure for the study or layout of the controls and displays. Participants were referred to the user guides provided if they asked any questions about the function or operation of control devices. At the end of 5 minutes, the experimenter stopped the practice session and asked the subject if they understood what they were required to do and clarified any confusion.

After the practice session, the simulation appropriate to the experimental condition was started by the experimenter. It ran for 22 minutes with a home-heating goal presented textually at the top of the screen every 2 minutes. The goals represented typical home-heating goals for a family with young children and were the same in each condition. To direct action, specific time frames, rooms or objectives were provided in the goal description. To signal a change of goal, the screen flashed yellow. On reading the goal, the participant was required to decide what adjustment of heating controls was necessary to achieve the goal, and perform any operation they thought appropriate (even if this resulted in no adjustment). If a subject had not completed their intended adjustments before the next goal was presented, they were to move onto adjustments for the new goal. At the end of the experiment, the screen flashed yellow and text (where the goals had formally been presented) informed the subject that the simulation was over.

Once the paper-based questionnaires had been completed, the experimenter turned off the TV monitor and removed the questionnaires and user guides to prevent them acting as prompts to the structured interview. The experimenter sat to the right of the subject at the same desk to conduct the structured interview. The audio recorder was switched on and placed on the desk. The experimenter explained the structure and purpose of the interview using the instructions for interviewer provided on QuACk. During part 1 of QuACk, the participants' answers to questions about their background experience with home-heating systems and attitudes to home-heating use were recorded by pen on the interview script by the experimenter. The participants' preferred terminology for home-heating components were written on individual post-it notes in marker pen. During part 3 of QuACk, the experimenter made clear to the subject that they should answer questions based on their experience with the simulation, not their own home-heating system. Participants were told they could refer to their own heating system as a means of comparison (e.g. 'like my heating system, it worked like....' or 'unlike my thermostat at home, it worked like....') but only descriptions relating to the simulation would be represented. Following the interview prompts, a diagram was constructed on the A3 paper by positioning post-it notes and linking and annotating by pen to represent the participants' UMM of the heating system presented on the simulation. On completion of this diagram, the experimenter paraphrased the mental model description to check for understanding and provide an opportunity for amendments. When any amendments were complete, a verification stage was undertaken, whereby each element, link and rule on the diagram was considered in turn and the subject asked to identify how confident they were that this represented what they believed. On completion of the verification stage, the subject was debriefed and paid £10 for their participation.

8.3   RESULTS

This section describes data gathered from UMM descriptions, variables relating to their behaviour with heating controls in the simulation and goal attainment based on room temperatures. The presentation of information will be split into three sections relating to these areas. In each section, the key hypotheses will be explored by tabulating descriptive statistics, then applying the Mann–Whitney U test for non-parametric data to determine the significance of differences in the realistic and design groups. Non-parametric tests were necessary as the data were not normally distributed (Field 2000). To drill into the detail of any significant differences, the data were illustrated in graphs and diagrams. Where graphs revealed the likely focus of differences, the Pearson's chi-square for categorical data was used to check the statistical significance. Where the data did not meet the criteria for Pearson's chi-square (i.e. expected values per cell were less than 5), the Fisher's exact test was applied.

8.3.1   USER MENTAL MODELS OF HOME-HEATING SIMULATION

Hypothesis A_1: Greater range of home-heating controls present in participants' UMMs when exposed to the design condition.

To determine if the design interface increased the discoverability of the home-heating controls, hypothesis A_1 predicted that the design condition would promote a greater

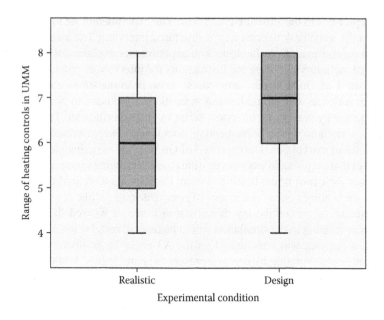

FIGURE 8.4   Range of heating controls found in user mental models.

range of heating controls in participants' UMM descriptions. Figure 8.4 shows a boxplot illustrating the median, interquartile range and minimum and maximum of the number of controls described by participants in their UMM descriptions. Using the Mann–Whitney test, it was found that the number of heating controls captured in participants' UMMs was significantly greater in the design condition than in the realistic condition ($U = 124.5$, $Z = 2.092$, $p < 0.05$, $r = -0.33$). The design condition therefore encouraged participants to include a greater range of heating controls, supporting hypothesis A_1.

Figures 8.5 and 8.6 group data indicating the presence of heating controls in UMMs by 'key controls' and 'advanced controls'. Whilst Figure 8.5 revealed little difference in the discoverability of key controls between conditions, greater variation was seen in number and type of advanced controls present in UMMs (Figure 8.6). Chi-square tests revealed a significantly greater use of the 'holiday button' ($\chi^2 = 10.99$, d.f. $= 1$, $p < 0.001$) and 'frost protection' ($\chi^2 = 32.40$, d.f. $= 1$, $p < 0.0001$) controls in the design condition.

Hypothesis A_2: Improved functional models of key devices are held by participants in the design condition.

Hypothesis A_2 predicted that participants in the design condition would hold more appropriate functional models of key devices than those in the realistic condition. Figure 8.7 shows a boxplot illustrating the median, interquartile range and minimum and maximum of appropriate functions described by participants to key controls in their UMM diagrams. Results of the Mann–Whitney U test found the number

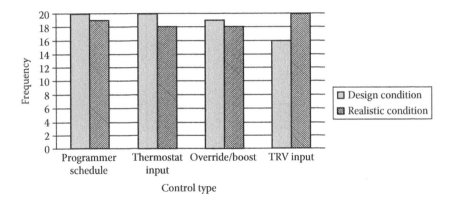

FIGURE 8.5 Frequency of participants who described key heating controls in user mental model descriptions.

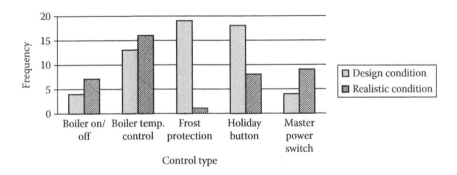

FIGURE 8.6 Frequency of participants who described advanced heating controls in user mental model description.

of appropriate functional models of key controls captured in participants UMMs was significantly greater in the design condition than in the realistic condition (U = 108.00, Z = −2.617, p < 0.01). To take into account variations in the number of key controls present in UMMs, a chi-square test was also performed comparing appropriate and inappropriate models, revealing a statistically significant difference (χ^2 = 7.335362, d.f. = 1, p < 0.01). Both tests support hypothesis A_2, that participants in the design condition were more likely to have an appropriate functional model of key devices.

The graph in Figure 8.8 illustrates differences in the appropriateness of functional models for the different controls described in UMMs. This shows that the programmer schedule is functionally understood by all participants, regardless of condition. In this sample, more participants in the design condition held an appropriate functional model for the boost and thermostat controls, but in both conditions,

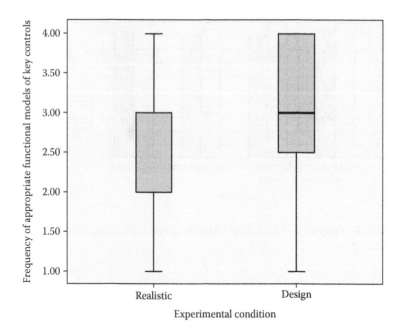

FIGURE 8.7 Frequency of appropriate functional models for key controls.

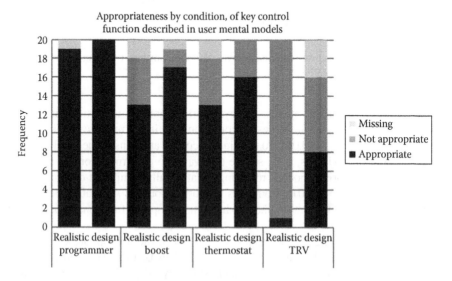

FIGURE 8.8 Graph to compare the frequency of appropriate and inappropriate functions assigned to key controls.

the majority had an appropriate model. Figure 8.8 shows that a statistically signifi-cant difference was found in the functional model held for the TRV control (χ^2 = 9.60, d.f. = 1, p < 0.01). Almost half of participants in the design condition had an appropriate device model of the TRV however, compared to a single participant in the realistic condition.

Hypothesis A_3: Improved number of key home-heating system elements described in UMMs of participants in the design condition.

Hypothesis A_3 predicted that participants from the design condition would describe a greater number of key home-heating system elements. Figure 8.9 shows a boxplot illustrating the median, interquartile range and minimum and maximum of key sys-tem elements described by participants in their UMM diagrams. It was found, using the Mann–Whitney test, that the number of key system elements present in UMMs was significantly greater in the design condition than in the realistic condition (U = 124.5, Z = 2.092, p < 0.05, r = −0.33), supporting hypothesis A_3.

The graph in Figure 8.10 compares the frequency of each key element found in UMMs. The largest differences relate to increases in design condition in the presence of the 'conditional rule', 'TRV feedback link' and TRV active indicator. Chi-square tests showed these differences were significant for the conditional rule (χ^2 = 5.226667, d.f. = 1, p < 0.05) and the TRV feedback link (χ^2 = 7.025090, d.f. = 1, p < 0.01). Fisher's exact test showed a significant difference for presence of the TRV active indicator in UMMs (p < 0.01). This indicates the design condition was more effective at encouraging increases in the presence of these elements.

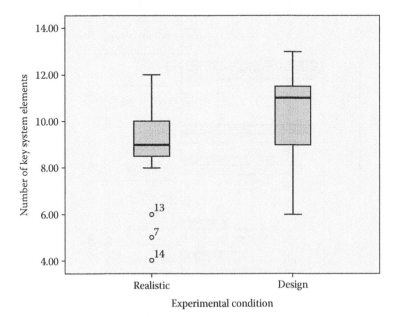

FIGURE 8.9   Number of key system elements present in UMM descriptions.

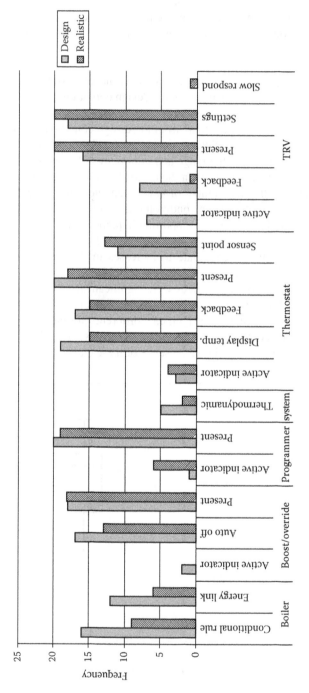

FIGURE 8.10 Frequency of key system elements present in UMM descriptions.

8.3.2   User Behaviour with Home-Heating Simulation

Hypotheses B_1 and B_2 predicted that participants in the design condition would adopt more appropriate behaviour strategies in line with the content of UMMs. For this study, significant differences were found between conditions with the prevalence of the frost protection and holiday buttons devices in UMMs. Hypothesis B_1 therefore predicts that the (i) frost protection and (ii) holiday buttons are included in more behaviour strategies in the design condition than in the realistic condition. Significant increases in the appropriateness of the functional model (at the device level) of the TRV and inclusion of the TRV feedback link (at the system level) were found in the design condition. Finally, a significant increase in the occurrence of the 'conditional rule' at the system level was found in the UMMs of participants in the design condition. Understanding the conditional rule enables deliberate control of boiler activation. Hypothesis B_2 therefore predicts that (i) the TRVs in the design condition will be operated in a way consistent with a temperature-sensing feedback device, and that (ii) effective boiler control would occur more in the design condition.

8.3.2.1   Underlying Assumption for Study

This study is underpinned by an assumption that if a control is present in a UMM description, this control is available for participants to include in behaviour strategies. This underlying assumption predicted that if a heating control is present in a UMM, it will be present (where the goal requires), in a home-heating behaviour strategy. Chi-square tests revealed a highly significant difference for both the design condition ($\chi^2 = 78.268$, d.f. $= 1$, $p < 0.0001$) and the realistic condition ($\chi^2 = 90.496$, d.f. $= 1$, $p < 0.0001$). For the design condition, 133 control elements were present in UMMs and 90.2% of these controls were used during the simulation; 47 controls were absent from UMMs, of which 76.6% were also absent in participants' behaviour in the simulation. The same trend was found in the realistic condition, with 89.7% of 116 controls present in UMMs being used in the simulation. Similarly, of the 111 controls absent from the UMMs, 77.9% were also missing from behaviour strategies (see Figure 8.11).This supports the assumption that contents of UMM descriptions captured by QuACK are linked to behaviour with home-heating controls.

Hypotheses $B_1(i)$ and (ii): Differences in the inclusion of specific devices in behaviour strategies.

The graph in Figure 8.12 shows that the majority of all participants used all four key controls (programmer, thermostat, boost and TRV). It also reveals a considerable difference in use of the frost protection and holiday buttons. No participants in the realistic condition used the frost control button, compared to almost all participants in the design condition. Chi-square tests showed these differences were statistically significant for the frost control button ($\chi^2 = 36.190$, d.f. $= 1$, $p < 0.0001$) and for the holiday button ($\chi^2 = 7.619$, d.f. $= 1$, $p < 0.01$), supporting hypotheses $B_1(i)$ and (ii).

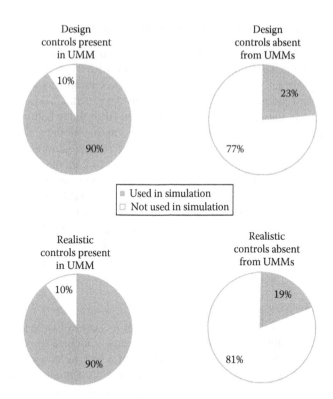

FIGURE 8.11   Proportion of controls used in simulation, depending on presence in UMM.

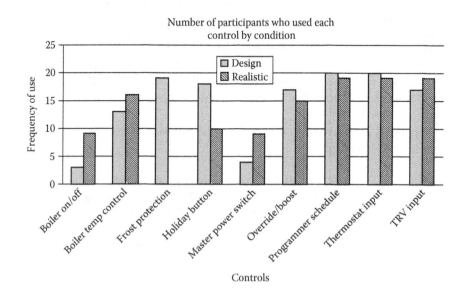

FIGURE 8.12   The frequency of use for controls.

Hypothesis B$_2$: The adoption of more appropriate set point values and frequency of adjustment for specific devices.

Hypothesis B$_2$(i): TRV operation consistent with a temperature-sensing feedback device.

Hypothesis B$_2$(i) predicted that TRV operations in the design condition will be operated consistent with a temperature-sensing feedback device. Less frequent and less extreme set point adjustments are more consistent with appropriate operation of a temperature-sensing feedback device for typical home-heating goals (Kempton 1986).

Figures 8.13 and 8.14 show boxplots illustrating the median, interquartile range and minimum and maximum of the total frequency of TRV set point adjustments, and mean range of TRV set point choices, respectively. Performing a Mann–Whitney test for non-parametric data failed to reveal a significant result (U = 158.000000, Z = −1.137224, p = not significant) for frequency of adjustment, but showed a statistically significant difference in the range of TRV set points (U = 110.500, Z = −2.428742, p < 0.05).

Figure 8.15 shows the adjustment strategies of TRVs for each condition. The set point choices in the realistic condition are more extreme and vary in direction far more than in the design condition, which shows a more subdued pattern. The greatest variation can be seen with the lounge, kitchen and children's bedroom, reflecting the target rooms in the majority of the provided goals.

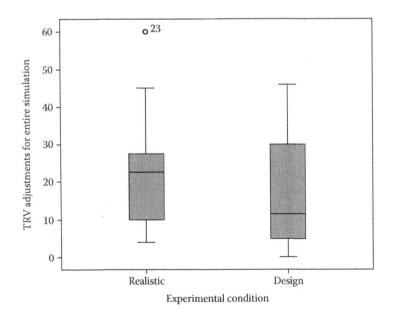

FIGURE 8.13   Frequency of TRV set point adjustments.

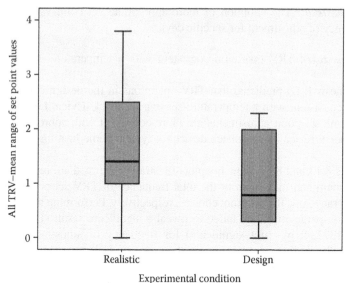

FIGURE 8.14   Mean range of TRV set point values.

Hypothesis B₂(ii): Differences in control of boiler activation.

Hypothesis B_2(ii) predicts that effective boiler control would occur more in the design condition. An essential prerequisite for boiler activation is for the thermostat to hold a higher set point than the hall room temperature, and a lower set point for deactivation. To test the statistical significance of this hypothesis, an independent samples t-test for parametric data was performed to compare the percentage of thermostat set point value changes that crossed the current hall temperature value. The results showed a statistically significant increase in control of boiler activation in the design condition ($t = 3.296$, d.f. $= 37$, $p < 0.01$), than in the realistic condition, supporting *hypothesis B_2(ii) (Figures 8.16 and 8.17).

Hypothesis C: Data relating to goal achievement through target temperature durations.

Goal achievement was based on target rooms achieving room temperatures within a target temperature range during a target time period. Where the target related to multiple rooms, the median room was used as the basis for measuring the duration of goal achievement as it reflected central tendency for non-normally distributed data. As target goal durations differed, to prevent this becoming a confounding variable, the proportion of time each goal was achieved was used. These were summed for 18 goals and converted into a percentage of overall goal achievement. Figure 8.18 shows boxplots illustrating the median, interquartile range and minimum and maximum of goal achievement. A Mann–Whitney test was undertaken, showing a statistically significant increase in goal achievement in the design condition ($U = 125.500$, $Z = -2.015$, $p < 0.05$), supporting hypothesis C.

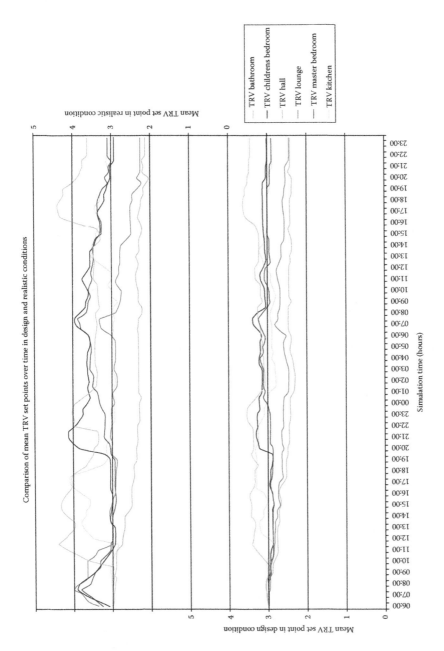

FIGURE 8.15 Frequency of use and set point choice over time of TRVs.

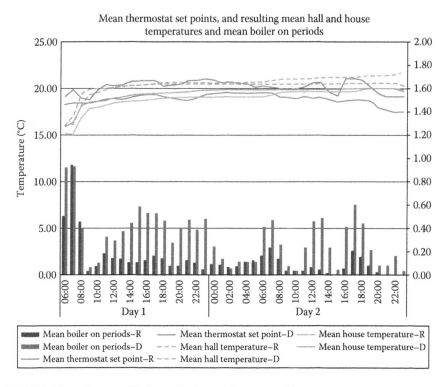

FIGURE 8.16 Control of boiler activation by thermostat adjustments.

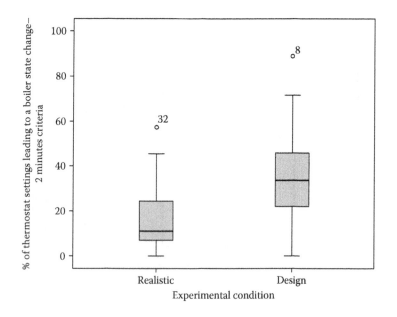

FIGURE 8.17 Percentage of thermostat set point choices leading to boiler state change.

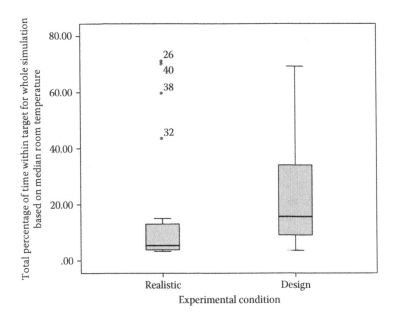

FIGURE 8.18 Total proportion of time within goal temperature range.

8.4 DISCUSSION

The underlying assumption for this study derived from Kempton (1986), and extended by Revell and Stanton (2014) in Chapter 3, is that householders' UMMs of their home-heating system affect their behaviour with that system. This underlying assumption was supported by comparing the presence of heating controls in participants' UMMs together with at least one instance of adjustment during the experiment. The statistically significant difference between the use of controls based on presence in UMM descriptions was not reliant on the type of interface presented, as comparable results were found for both conditions (see Figure 8.11). Identification of a link between UMMS and behaviour complements previous literature from other domains (Gentner and Stephens 1983, Kieras and Bovair 1984, Wickens 1984, Hanisch et al. 1991). The QuACk method for capturing UMMs is proposed as a useful tool in understanding and predicting whether a device is likely to be used. The level and appropriateness of use is subject to more detailed analysis, however, and the subsequent hypotheses go some way in exploring the value of the method in that respect.

8.4.1 IMPROVED DISCOVERABILITY OF HOME-HEATING CONTROLS

The greater range of heating controls present in UMMs of participants in the design condition (supporting hypothesis A_1) suggests that the changes in the home-heating interface improved the discoverability of controls overall (see Figure 8.4). The difference in discoverability of home-heating controls in the realistic condition supports findings by Shipworth et al. (2009), and Revell and Stanton (2014) in Chapter 3.

When polling 427 English homes, Shipworth et al. (2009) found many people did not recognize many of their home-heating controls. In Chapter 3, Revell and Stanton (2014) found key heating controls were missing from UMMs of occupants new to heating control. Similarly, Brown and Cole (2009), who compared heating controls in standard and 'green' office buildings, found that the highest reason occupants had for not using controls was 'Controls don't exist' followed by 'I don't know where they are'. The implications of a greater number of controls described in UMM descriptions in the design condition are that the appropriate control(s) are available in the UMM to draw upon when determining an action specification to fulfil a given goal (Norman 1986). However, it should be noted that since both conditions had a high prevalence of the key heating controls (see Figure 8.5), the differences in the inclusion of controls in behaviour strategies, attributable to discoverability alone, would be most marked when attempting to fulfil goals met by 'advanced controls' (in this case, the frost protection and holiday buttons). Improvements in discoverability of less familiar controls by representing physically distributed devices in a single 'control panel' could hold the key to fulfilling the potential of energy-saving systems and monitoring technology (e.g. smart meters). Glad's (2012) study on the benefits of new energy systems in Swedish housing cited that lack of discoverability of devices thwarted expected improvements in CO_2 saving.

8.4.2 More Appropriate Mental Models

Participants from the design condition were found to have more appropriate functional models of key heating controls, and key system elements supporting hypothesis A_2 (users were more likely to hold an appropriate functional model of key devices) and hypothesis A_3 (users would describe a greater number of key mental model elements), respectively (see Figures 8.7 and 8.9). These results are particularly encouraging, given the lack of formal 'training' to promote specific model types in comparison to studies like Kieras and Bovair (1984) who provided extensive training, followed by a test for comprehension and a re-test a week later to ensure knowledge retention, and Halasz and Moran (1983) who provided 30 minutes training in advance of their experiment. Both of those studies used novice participants who did not have to 'overcome' an existing mental model of the test device. This study, in comparison, comprised of participants with between 5 and 40 years' experience of home-heating controls, so existing knowledge structures would need amendment (Johnson-Laird 1983). The largest difference between conditions was found with the functional model of the TRV as a feedback device and the conditional rule for the boiler, which requires both the thermostat and the programmer (or boost control) to be 'calling for heat' to trigger boiler activation. The function of the TRV was misunderstood by most participants, supporting the findings found in Chapter 6. Improvements seen in the appropriateness of the TRV functional model backs the view by Kempton (1986) that when the operation of controls is 'visible', the correct model is adopted. Improvements in the presence of the 'conditional rule' element reflect advice from Kieras and Bovair (1984:271) that the most useful information to provide users is 'specific items of system topology that relate the controls to the component's and possible paths of power flow'. Contrary to the work by Kempton (1986), Norman (2002) and Peffer (2011), but in support of Revell and Stanton (2014) described in Chapter 3, most participants provided an appropriate functional model for the thermostat device (see Figure 8.8) suggesting present-day UK householders *do* understand this device. The programmer control was not only included in all but one of participants' UMMs, but its purpose was well understood. No inappropriate functional models were given by participants for this device, so issues with operation are likely to result from inappropriate models at a system level, or the known usability issues with typical device designs seen in the literature (e.g. Combe et al. 2011, Peffer et al. 2011). That amendments to existing UMMs of home-heating systems can be achieved within a very short period of time (25 minutes of accelerated interaction) without 'formal instruction' has favourable implications for using UMM-based design for home-heating systems. This is likely to be generalizable to other systems where inappropriate UMMs have been shown to result in inappropriate behaviour.

8.4.3 INCREASED USE OF FROST PROTECTION AND HOLIDAY BUTTONS

Hypothesis B_1 represented differences in behaviour with heating controls following from statistically significant differences in UMM content. At the device level, this focused on difference by condition in the use of the frost protection and holiday buttons. As expected, the design condition encouraged significantly more interaction with these controls than the realistic condition. Whilst the former device is concerned with safety, increased use of the holiday button (that replaces the existing programmer schedule with minimal on periods to reflect lack of occupancy) is clearly relevant in terms of reducing energy consumption. How significant this is in terms of other recommended behaviour requires further research. That lesser known controls are used when present in UMMs, and accessible on an interface, lends support to the view that increasing the discoverability of energy-saving technology could indeed increase inclusion in behaviour strategies, potentially realizing the energy-saving potential promised by technology. This would, however, still rely on the adoption of corresponding energy-saving goals (Norman 1983, Moray 1990, Bainbridge 1992, Chapter 6).

8.4.4 MORE APPROPRIATE BEHAVIOUR WITH TRV CONTROLS

Following from analysis of the appropriateness of functional models of key control devices, corresponding differences by condition in the behaviour patterns relating to TRVs were expected (hypothesis $B_2(i)$). Figure 8.15 showed TRV adjustments over time for both conditions and it was clear that expert advice promoting static set points (Revell and Stanton 2016) were not evident in either condition. However, this can be explained by the style of the prescribed goals for 'comfortable' conditions in specific rooms, which may have convinced participants that this level of custom control was possible, as well as the lower effort levels necessary to make changes to distributed controls when presented on a single interface. The results in this study showed increased interaction with TRVs in the realistic condition, but this was not statistically significant (see Figure 8.13). A statistically significant difference was found, though, in the mean range of set point adjustments (see Figure 8.14).

The more extreme set point choices found in the realistic condition are reminiscent of the behaviour pattern described by Kempton (1986) for 'valve' model holders of a central thermostat, whilst the moderate adjustments made in the design condition are more in line with Kempton's (1986) behaviour pattern corresponding to a 'feedback' model. Further examination of the data revealed that a 'valve' functional model for the TRV was held by the majority of participants in the realistic condition. This supports the view given in Chapter 4 and by Revell and Stanton (2014) that Kempton's insight can be applied to alternate control devices, and suggests that differences in UMMs in terms of function can explain behaviour patterns. Whilst TRV is not cited in the literature as a key player affecting domestic energy consumption, issues of unnecessary adjustment may not be considered a problem at the device level. However, when viewing heating controls as an integrated system, Chapter 5 (Revell and Stanton 2015) showed that inappropriate UMMs of other controls, such as the TRV, can have a significant negative impact on both comfort and energy consumption.

8.4.5 Greater Control of Boiler Activation

Hypothesis B_2(ii) examined differences in boiler activation predicted from significant differences in the presence of the 'conditional rule' in UMMs. To intentionally fulfil heating goals and manage energy consumption, it is necessary for the participant to have an understanding of the link between the set points of the thermostat, its relationship to sensed temperature in order to 'call for heat' and its dependency on the setting of the programmer and boost for boiler activation. The graph in Figure 8.16 shows the mean number of thermostat adjustments over time for both conditions. Whilst a static thermostat set point is encouraged in manuals and by expert advice (Revell and Stanton 2016), the appropriate set point choice requires an appreciation of the thermodynamics of the house structure. Crossman and Cooke (1974) emphasized that operator control of dynamic systems requires sufficient time for experiment and observation, which was not provided to participants in this experiment. In addition, the (unintended) non-typical thermodynamic model for the simulation resulted in particularly high temperatures in the hall where the central thermostat was located. This meant that far higher set point values would be necessary to activate the boiler than participants would be used to selecting at home. Expectations for a static thermostat pattern were unreasonable in these circumstances. The mean data represented in Figure 8.15, suggest that participants in the design condition were able to influence boiler activation by varying the thermostat set point (thick orange line) above and below the hall temperature value (dashed orange line). Participants in the realistic condition display very little influence over boiler activation as after 0800, day 1, the mean thermostat set point (thick blue line) remains below the mean hall temperature (dashed blue line). For the majority of participants in the realistic condition, therefore, the boiler would be inactive for a substantial part of the simulation, despite repeated set point adjustments in both directions. Results from a t-test supported the hypothesis B_2(ii) that participants in the design condition operated a greater level of intentional control over the boiler. This result supports the findings of Kieras and Bovair (1984) who found that participants with an appropriate UMM of the system engaged in very few 'nonsense' actions, favouring behaviours that were consistent with the device model. This result further supports what Revell and Stanton (2014) described in Chapter 3, that UMMs of home heating must be considered at the system level, since the majority of participants in the realistic condition had an appropriate functional UMM for the thermostat device in isolation. Usability or mental model promoting design initiatives focused solely at the single device level (e.g. programmer, thermostat, TRV) are unlikely to maximize desired benefits for comfort or consumption goals.

8.4.6 INCREASED GOAL ACHIEVEMENT

Participants in the design condition were also significantly more successful at achieving the goals provided (hypothesis C). This result supports the work of Kieras and Bovair (1984) and Hanisch et al. (1991). To be as realistic as possible, comfort goals made up a large part of goals, and energy conservation was the focus only when the house was to be unoccupied. The proportion of goal achievement was relatively low in both conditions compared to the results of Sauer et al. (2009), who found the proportion to be between 73% and 94%. Sauer et al.'s (2009) study tasked participants with choosing setting in advance to achieve a specific daily profile, allowing greater opportunity for planning and amendment. In contrast, this study presented a succession of changing goals that incorporated not only typical planned changes in comfort goals, but a more realistic 'ad hoc' adjustment of goals throughout the day, which were not necessarily achievable. This study also placed participants with greater time pressure to make their heating adjustments. Previous work in Chapters 3 and 5 has led to the conclusion that a 'gap' between heating expectations (in terms of speed of achieving comfort goals, consistency through the house in space heating temperature and effectiveness of 'custom' room heating) and what typical UK heating systems can deliver has a large influence on encouraging inappropriate behaviour. Greater generalizability of the results to everyday behaviour is therefore possible by providing 'realistic' goals in the simulation. The measure for goal achievement was based on temperature values for rooms rather than consumption or deployment of appropriate action sequence with controls. This meant that appropriate choices in behaviour (e.g. programmer settings, or deployment of the frost protection and holiday buttons) did not get recognized unless there was an impact on room temperature values within the target time period. Further analysis that matches behaviour strategies to specific goals will be the focus of further work. The most important aspect of this study is that performance improvements can be explained by better control of boiler activation following from increased understanding of the conditional rule, resulting from design changes in the interface and instruction manual. This result is highly encouraging as it demonstrates that home-heating performance can be affected through design for a given set of goals.

8.4.6.1 Limitations of Study

There were limitations of the thermodynamic model used for the simulation. The mean house temperature over time for each condition shows a gradual increase in value throughout the simulation (see Figure 8.16). Given the variations in the controls used and boiler on periods, this suggests the insulation parameters were unusually high, preventing heat loss from occurring. This has implications in terms of energy-saving goal achievement based on room temperature, since a deliberate drop in room temperature cannot be contrived by behaviour strategy. The thermodynamic characteristics of the simulation were therefore atypical and limit the generalizability of goal achievement findings, particularly relating to comparison in energy conservation. To reduce the variables under analysis, the simulation did not allow the participant to control ventilation or heat flow within the house through opening and closing of doors and windows. Comments from participants indicated that this would have made up part of their strategy for controlling room temperatures, suggesting that restricting behaviour to heating controls only may not be representative for how people manage space heating in the home.

8.5 CONCLUSIONS

Differences in the design of an interface have been shown to change the content of mental model descriptions. The hypotheses that correspond with the design condition to improve the functional mental model of heating controls, the discoverability of heating controls and the number of key system elements were supported. Key differences in users' mental model descriptions focused on the TRV control (appropriate model and feedback link), an awareness of the conditional rule for the boiler and the presence of the frost control and holiday buttons. Differences in the content of mental model descriptions were found to correspond with differences in operation of the heating simulation. Inclusion of controls in a mental model was a highly significant predictor of whether that control would form part of a behaviour strategy. The action specification with controls was also seen to vary between conditions in line with the key differences found in users' mental models, significantly improving TRV set point choices and control of boiler activation by the central thermostat in the design condition. Participants in the design condition achieved significantly greater proportion of the experimental goals than those in the realistic condition. It is concluded that a control panel interface that promotes a UMM integrating heating controls with energy monitoring and predictive technology will help users to have more control over heating consumption. This simulation focused on communicating control device models and interdependency of devices. The promotion of appropriate UMMs of home space heating within the broader context of thermodynamics, weather conditions and house structure would be the natural next step to continue this research.

9 Conclusion

9.1 INTRODUCTION

The focus of this book was to investigate how the concept of mental models could be applied to interface design to encourage behaviour change with domestic energy systems. The domestic heating system was presented as an in-depth case study, as this makes the largest contribution to energy consumption in the home. Many of the behaviour trends, device models and design solutions would be equally applicable to domestic cooling systems, given their similar goal to ensure a comfortable temperature in the home. Indeed, given the projections for temperature rise post 2050 due to climate change, the application to cooling systems may well be increasingly pertinent. The approach demonstrated, however, is generalizable to any system where its function is not explicit, yet people are required to make cause-and-effect judgements to set up and operate its controls. Whilst the overriding impetus of the research was to reduce consumption, it is the relationship between householders' mental models of the home-heating system and resulting behaviour that has necessitated focus. The key findings are summarized herein, followed by a discussion of the core issues underlying the research. Finally, key recommendations and areas for future work are presented.

9.2 SUMMARY OF FINDINGS

This book explored Kempton's (1986) insight that folk models of the home-heating thermostat were associated with energy-wasting behaviour patterns. Considering bias in its development, a semi-structured interview was constructed to elicit, describe and analyse householders' mental models of home-heating systems and self-reported behaviour. It was found that present-day householders used a range of strategies including multiple controls to manage home heating. Considering mental models and behaviour patterns of thermostats in isolation was no longer appropriate, a systems view was proposed to understand domestic home-heating behaviour and its effect on consumption. Analysis of mental models and behaviour strategies at the level of the whole heating system revealed gaps and misunderstandings that hindered appropriate behaviour. Differences in householders' goals also affected the behavioural strategies adopted. A design specification was developed by comparing the gaps and misunderstandings in novices' mental models of the home-heating system to that of an expert. For optimal consumption, it was recommended that householders incorporate in their mental model, broader variables that impact consumption rates. A study comparing a control panel interface developed to promote a compatible user mental model (CUMM) with a more typical interface found that the more users described more appropriate mental models at the device and system levels, the greater was the control of boiler activation, and greater was the goal achievement.

9.2.1   BIAS MUST BE CONSIDERED IN MENTAL MODELS RESEARCH

As mental models cannot be accessed directly, methods for eliciting, describing or interpreting related data are subject to bias. Confidence in application of findings is limited unless the impact of bias is made explicit. A method for considering bias in mental model research was not evident in the literature, so the 'tree-rings' framework was developed in Chapter 2. This method allowed systematic consideration of bias in definition, capture, analysis and representation of mental models. This method enables researchers to graphically represent the relationship between the mental model source, intermediaries and recipients when conducting mental models research. Analysts are then prompted to consider the background, social and cognitive artefact biases that come into play. By translating these diagrams to 'tree-ring' diagrams, key characteristics of the type of knowledge structure captured, and the layers of bias to which it is subject could be visualized. This technique is a domain-independent tool that can aid researchers when constructing methods or undertaking studies involving knowledge structures. The tree-ring method also offers a novel way of gauging the commensurability of existing mental models research.

9.2.2   OUTPUTS FROM QuACk HELP EXPLAIN ENERGY-CONSUMING BEHAVIOUR

The tree-rings framework was used to develop a systematic method to capture user mental models of home heating and associated behaviour. The Quick Association ChecK (QuACk) is a semi-structured interview script developed with consideration of bias for data collection, representation and analysis. QuACk was applied in this book, first to 3 pilot participants (Chapter 3), then to 6 householders in matched accommodation (Chapters 4 through 6), as well as 40 participants of a home-heating simulation (Chapter 8), and to 1 expert of home-heating controls (Chapter 6). Evidence of differences in user mental models of the thermostat and other control devices were found, extending the work of Kempton (1986) (Chapters 3 through 5 and 8). The shared theories (valve, feedback, timer, switch) described in the existing literature (Kempton 1986, Norman 2002, Peffer 2011) on the thermostat device were useful in a 'generic' capacity to describe functional models and behaviour for alternate devices (Chapters 3, 4 and 8). Self-reported behaviour with home-heating controls (depicted in graphical form using a QuACk template) was helpful for understanding variations in users' energy-consuming behaviour strategies (Chapters 5 and 6). The content of mental model descriptions explained differences in householders' behaviour with heating controls. Omissions of control devices in users' mental model descriptions explained omissions in users' self-reported behaviour (Chapters 3 through 5) and actual behaviour in a home-heating simulation (Chapter 8). Differences in the functional models of the thermostat control explained differences in the way householders reported operating the device (Chapter 4). Similar results were found for the TRV controls and actual behaviour shown in the home-heating simulation. Householders with the same functional model of the thermostat, but differences in the range and functional models of other home heating devices, showed considerable variations in their reported behaviour strategies and recorded consumption levels (Chapter 5).

Differences between the content of householders' mental model descriptions and that from a home-heating expert, highlighted common misunderstandings and omissions relevant for targeting energy-saving strategies. The method for categorizing outputs has been validated by two human factors analysts (Chapter 3).

9.2.3 WE NEED TO THINK BEYOND THE THERMOSTAT – HOME-HEATING BEHAVIOUR SHOULD BE UNDERSTOOD AT A SYSTEM LEVEL

The focus of past literature on mental models of home heating and associated behaviour focused on the thermostat device. Chapters 3 through 5 revealed other devices were used as the main control (e.g. programmer, on/off switch or TRVs), resulting in static or very infrequent thermostat set point adjustments. To understand the home-heating behaviour strategy adopted by a householder, set point adjustments need to be viewed for the whole heating system, not just a single device. Home heating controls are integrated in function, so set points on one device affect operation of other devices (cooling system operate in a similar way). Mental model descriptions by householders and participants revealed misunderstandings about how controls are integrated (Chapters 3 through 5 and 8). These misunderstandings explained non-optimal operation of controls by householders, leading to compromised comfort (Chapters 3 and 8) or wasted consumption. Other domestic energy systems that rely on integrated controls could equally compromise their key benefit and consumption if a system-level approach to interface design is not adopted.

9.2.4 BROADER SYSTEM VARIABLES NEED TO BE UNDERSTOOD FOR OPTIMAL CONSUMPTION, BUT ARE NOT PROMOTED BY EXISTING TECHNOLOGY

To balance comfort and consumption with home-heating controls, expert recommendations revealed that users' mental models needed to include concepts relating to the context of use (Chapter 6). To select an appropriate thermostat set point, householders need to be aware of variations in comfort levels throughout their home. To select appropriate programmer start and end times, they need to be aware of, and able to quantify, lag times in heating and cooling. To appreciate energy consumption over a particular time period is greater at night, or when internal doors are open, an understanding of how household thermodynamics are affected by infiltration of warm air throughout the home, and temperature differentials, is needed (Chapter 5). These types of concepts extend beyond the heating system as they relate to variables associated with house structure and weather conditions. Kempton (1986) highlighted how the feedback 'folk model' of the thermostat could result in wasted energy because it did not incorporate these broader variables. Typical home-heating technology in the United Kingdom does not make this visible nor communicate the influence of these variables on comfort or consumption (Chapter 6), hindering the formation of CUMMs for optimal home heating. When approaching mental model interface design for other domestic energy systems, it is key to understand the variables beyond the control of the user and technology for realistic expectations of impact.

9.2.5 MENTAL-MODEL-DRIVEN DESIGN HELPS USERS ACHIEVE MORE HEATING GOALS

Norman's (1983) seven stages of activity, and gulf of execution and evaluation were applied to the home-heating context. This resulted in a design specification to promote a CUMM that enabled recommended operation of the home-heating system (Chapter 6). This design specification was used to develop a concept for a home-heating control panel that (1) improved discoverability of all controls, (2) improved the functional model for key controls, (3) promoted the interdependence of key controls, (4) highlighted the influence of control set point choices on radiator output and boiler activation and (5) provided real-time thermodynamic feedback. Compared to a more traditional design, the control panel design significantly improved the understanding of the range of heating devices, the appropriateness of their functional model and understanding of the way key devices were integrated in users' mental models. Significantly greater control over boiler activation and achievement of home-heating goals was observed with the control panel design. Whilst normally adopted in industry, for domestic energy systems with multiple integrated controls, a control panel interface provides a design solution that enables a 'retrofit' compatible with existing technology.

9.3 CORE ISSUES RELATING PARTICULARLY TO THE HOME-HEATING CASE STUDY

9.3.1 OPTIMAL HOME HEAT CONTROL IS A COMPLEX TASK

Sauer et al. (2009) described central heating as the most complex system in the domestic domain. Mental models are accessed when users try to interpret complex systems (Moray 1990). The central heating system is a slow responding system and, inherent in systems of this type, cause and effect is difficult to gauge by observation alone (Crossman and Cook 1974). In addition, the user is faced with multiple distributed controls that vary between households in their location, interface and functionality. Optimal comfort and consumption levels are dependent on the compatible adjustment of integrated controls. Additionally, they are affected by variables relating to the environmental setting. These include static variables such as house structure and level of insulation, as well as changing variables within the control of the user (infiltration due to door and window positions) and outside the control of the user (external temperatures varying throughout the day and over changing seasons). For optimal heating control, users have a number of different levels of understanding to navigate: (1) awareness of controls and the correction functional model at the device level, (2) awareness of how controls are integrated at the system level; (3) awareness of house structure characteristics on comfort and consumption and (4) awareness of climate characteristics on consumption. But this is only one side of the story. For optimal heating control, householders also need to match this understanding to an understanding about their own lifestyle. This means that they need to have an appreciation of their occupancy levels and those of others within the household, as well as the different needs of different members of the household (e.g. greater comfort levels for vulnerable occupants, lower room temperatures when cooking, exercising or doing housework, night-time comfort for those studying late). This is no mean feat, and Sauer et al. (2009) found it was far more difficult to conserve energy with variable daily routines. Given the barriers to forming a CUMM, householders are faced with matching their unique lifestyle goals with the demands of home-heating control by referring to what is often an incomplete mental model of the system. It is therefore unsurprising that the behaviour specifications that result are far from optimal.

9.3.2   EXISTING TECHNOLOGY DOES NOT SUPPORT A 'SYSTEMS UMM' OF HOME HEATING

The home-heating systems typically found in UK homes were designed to provide 'space heating', rather than an 'optimal balance between comfort and consumption'. However, through rising fuel prices and environmental concerns, householders are being tasked with operating a system designed for one purpose to achieve another purpose. To achieve optimal control, householders need to hold mental models of technology at a system, not device, level. However, the modular nature of home-heating devices hinders this goal. The controls, radiators and boiler found in typical households are manufactured by different companies. The companies themselves do not know the range or type of controls that will integrate with their device and so can only provide generic guidance outside the feature of their product. Boilers vary in their efficiency, whilst controllers vary in their interface, features and location within the home. In the United Kingdom, heating systems are generally inherited, rather than chosen. Unless aware of certain controls through prior experience, it is not surprising that some householders are wholly unaware of their existence. Heating controls and components are also limited in the type of feedback they give to users. Generally, this is limited to the set point, status and, where appropriate, temperature measurements. Feedback that would allow an understanding of cause and effect of their behaviour, the functioning of the heating system as a whole and the impact on heating goals is missing. Even at the device functional level, user manuals are often verbose, difficult to understand and, if not lost, filed away never to be read. Whilst technology has been used as a tool to encourage energy-conserving behaviour in the home, there has been a tendency to focus at the device level. Programmable thermostats have been adopted to help save energy at night or when the house is unoccupied, but do not emphasize the conditional nature of set point choice for boiler function, neither do they facilitate users choosing appropriate start and end times that fit their lifestyle and the heat lags associated with their own house. Smartmeters and energy monitors have been introduced to make householders aware of their consumption, but these initiatives rely on the user being able to interpret data at a system level in order to specify an appropriate behaviour strategy. This is a tall order without a full CUMM and a highly routine lifestyle. By failing to help householders interpret these strategies at the 'system level', householders will refer to and amend their own mental models of their heating system to inform their behaviour strategy. Given the misunderstandings and omissions in user mental models of home-heating systems revealed in this book, strategies to reduce consumption positioned at the 'device' level are at risk of *compromising*, rather than *optimizing* the balance between comfort and consumption.

It is not surprising that the EnergyStar rating for programmable thermostats have been removed (Energy Saving Trust 2013) and the benefit of a SmartMeter roll out in the United Kingdom is being questioned (BBC 2014). Householders currently have the wrong tools for the job of optimal consumption.

9.3.3 We Cannot Control All the Variables That Affect Optimal Home Heating Control

There are a number of broader system variables that influence comfort and consumption levels including people's lifestyles, the climate and the structure and insulation levels of the house. Whilst householders have some control over their lifestyles, they may not be able to dictate the regularity of their occupancy if this is linked to work and school obligations and other commitments outside their home. Although they can make choices relating to internal and external infiltration of air around the home, and the installation of insulation, it is more difficult for householders to make adjustments to the structure of the house to better support thermodynamics and evenly distributed heat. Without moving some distance, householders have no control over the external world climate to which they are subjected. Householders will always be subject to variations that make it difficult to ensure comfort or avoid wasting energy. Similarly, there is a limit to what government initiatives can do to control these broader variables, although subsidized insulation and the building of 'greener' buildings are positive steps. This book has focused on the problem of influencing householders' *behaviour* with home-heating controls, rather than these outside variables. But it is important to appreciate how appropriate householders' behaviour with controls is subject to many other variables. A good choice of set point at one time of day is a poor choice at another time of day. A programmed set of times optimally fits a lifestyle one week, but wastes energy the next week. What is considered a comfortable temperature for occupants sitting and watching TV in the evening is different when doing the housework. The provision of 'one-size-fits-all' prescriptive advice on how to manage home-heating systems is therefore unrealistic.

9.4 RECOMMENDATIONS

9.4.1 Recognize the Complexity of the Task for Householders, When Embarking on Strategies to Reduce Home-Heating Consumption

The complexity of the task that householders face to optimize consumption needs to be recognized before effective guidance can be provided. Whether technology-driven guidance, or government campaigns, simplistic, generic advice is inappropriate given the variations affecting householders. Tailored guidance that takes into account differences in householders' lifestyles and the influence of broader variables is more likely to result in appropriate home-heating management. Device certification (e.g. EnergyStar) of devices that rely on the adoption of specific behaviour habits should be tested in the context of use to gain robust indications of energy savings.

9.4.2   USE SYSTEM-LEVEL STRATEGIES FOR ENCOURAGING APPROPRIATE HOME-HEATING CONSUMPTION

A systems-level approach is prevalent in human factors research in a wide range of domains. The findings of this book indicate that a systems approach to tackling energy-consuming systems in the domestic domain would be beneficial. Before embarking on a strategy targeted at the device level, an understanding of the interdependency of this device with other devices in the system is necessary. Ensuring the 'user' understands these dependencies is crucial for success. This applies not only for the redesign of existing home-heating controls (e.g. improving the usability of programmable thermostats), but also the introduction of new technologies designed to aid energy reductions; for example, the introduction of energy monitors with a strategy that effectively communicates to householders how consumption feedback relates to chosen settings of key controls in different circumstances. System-level strategies that go beyond the central heating-system controls to include broader variables are likely to be even more effective to householders, as they would promote a compatible user mental model that enables appropriate consumption. In addition, providing householders with a systems-level view that considers heat loss rates could have the 'knock-on' effect of making explicit the benefits of investing in low-tech improvements such as insulation and draft excluders. It may even positively influence behaviour by making explicit the effect of leaving doors and windows open for longer than necessary.

9.4.3   USE A MENTAL MODELS APPROACH WHEN SEEKING TO ENCOURAGE APPROPRIATE BEHAVIOUR IN COMPLEX SYSTEMS

A mental models approach to system control has the benefit of aiding learning and facilitating troubleshooting (Norman 1983). Where operators hold an appropriate mental model of a system, variations in their goals can be accommodated (Moray 1990). This in turn can facilitate appropriate behaviour when undertaking tasks. In the case of home heating or cooling, this could lead to systematic improvements in goal achievement. Householders whose goals include reducing waste (e.g. energy or money) could systematically reduce consumption. Helping users to hold an appropriate 'picture in the mind' of cause and effect at the point of set point adjustment is possible through design driven by mental models research.

9.4.4   DESIGN FUTURE HEATING SYSTEMS WITH OPTIMAL CONSUMPTION AS THE PRIMARY GOAL

Ultimately, the way in which legacy home-heating technology is presented to householders, is no longer 'fit for purpose'. Designers for devices, such as home heating controls, which currently rely heavily on human behaviour for energy efficiency, have a responsibility for enabling energy-efficient behaviour in the context of use. New home-heating technology needs to be designed so optimal consumption is its primary goal. To do so effectively, the broader system variables need to be taken into consideration in their design.

9.5   AREAS OF FUTURE RESEARCH

9.5.1   Extension of the 'Tree-Ring' Method for Considering Bias

The tree-ring method was applied in this book to home heating and a bank machine. Application to other contexts where knowledge structures are being explored would further validate its generalizability. Further population of the types of bias that are likely to act at different levels, and guidance of typical biases found in particular circumstances would help analysts mitigate for bias in their research, improving confidence in results. Assessment of a significant number of different studies involving knowledge structures using the tree-ring method would provide an opportunity to appreciate their commensurability.

9.5.2   Extension of the QuACk Method for Exploring Association with Mental Models and Behaviour

The QuACk method was developed for the home-heating context in this book; however, this approach has also been adopted in the domain of water management (Kalantzis et al. 2016) and by the authors in the aviation domain, providing evidence of its generalizability (with adjustment of questions to match the domain). Further adoption of this method to explore domestic energy would validate it usefulness in this domain, as well as enable studies of reliability and validity.

9.5.3   Tailored Guidance for Optimal Home-Heating Behaviour in Different Circumstances

Application of Norman's (1986) 'gulf of evaluation and execution' in Chapter 6 considered simplified recommendations for generic heating goals to understand where householders deviated from expectations. Tailored guidance for appropriate behaviour based on householders' specific goals, which allowed householders to 'preview' predicted consumption, could better support optimal consumption or encourage goals that better fit the installed heating (or cooling) system. Research focused on modelling optimal behaviour in different types of structures, with different environmental conditions, and heating-system setups against a variety of user goals would be highly beneficial. Further work that looks at how best to communicate the recommended behaviour changes at the point of action would be a considerable step towards providing highly effective guidance.

9.5.4   Enhancement to Home-Heating Control Panel and Testing in Domestic Setting

To upgrade the design of the home-heating control panel, different modes would present different controls during setup and for ad hoc adjustment. Formal user testing would be undertaken at the device and system levels. Advanced features such as predictive guidance for different circumstances would be the ultimate goal. The control panel would need to be tested in a real-life setting to determine if improvement to user mental models and achievement of home heating goals hold 'outside the lab'.

9.6   CONCLUDING REMARKS

This approach to home-heating control is highly pertinent given the level of consumption and potential environmental consequences. This research investigates a 'retrofit' to technology that is no longer fit for purpose as its purpose has changed from space heating for comfort, to energy conservation. Similar approaches could be adopted for other domestic energy systems, or domains where devices are designed for a benefit rather than conservation and require active judgements relating to broader systems for efficient operation (e.g. motor vehicles, aviation vehicles, air conditioning). As pervasive computing/intelligent computing advances in its ability to tailor home heating to peoples' goals and expectations, the need to actively manage heating control may become a thing of the past. The literature often cites human behaviour as a barrier to reduced consumption, but this is an unfair charge. The barrier to reduced consumption is the design of technology that does not enable optimal operation and not the user of it. This book has shown how advances could be made in understanding behaviour as well as bringing about behavioural change through interface design using mental models.

Appendix A: The Quick Association Check

This is the full version of the semi-structured interview developed in Chapter 3 and used in Chapters 4 through 6, and in a form amended for the simulation in Chapter 8. This script is inherently suitable to explore domestic cooling instead of heating (merely by substituting reference to 'heating' – and associated concepts – with 'cooling'). With adjustment of questions, this questionnaire can be utilized for other domains looking to compare mental models with behaviour. The interview can also be used in component form if user behaviour or mental model outputs are desired in isolation. Outputs can be analysed using the reference tables and guidance questions provided in Appendix B.

A.1 QuACk INSTRUCTIONS FOR INTERVIEWER

A.1.1 BACKGROUND

The QuACk interview script is separated into three areas:

1. Background experience in home heating/capture of participant's home-heating terminology – questions and probes to guide the positioning of the interview, and understanding of participant's responses.
2. Behaviour with home-heating system – questions and scenarios to collect data relating to devices used, and the set points chosen over time. These are then used to populate a diagram of self-reported behaviour.
3. Mental model of home-heating system – questions and probes to identify home-heating components, their function and the rules and relationships between components. These are used to build up a diagram describing the participant's 'mental model' of the home-heating system.

Each area will start with 'verbal positioning' to the participant. It is important that this is not missed out, even if it seems repetitive. Throughout the interview script, there will be instructions to the interviewer in bold to provide guidance. Depending on the answers to section 1, some questions may need to be skipped (e.g. if they do not use a particular device) or adjusted (e.g. the terminology used to describe a device, or based on demographic information). The interview should be in a relaxed conversational style, so the participant is allowed to continue a train of thought where it relates to the data sought, but is brought back to the questions if it goes off-track. Be aware that some participants, when discussing heating, will have a preference for discussing temperature settings in °F rather than °C.

A.1.2 Preparation

The following equipment is needed to conduct the QuACk for home heating:

- Participant information sheet
- Interview script
- Self-report template
- Audio recorder and batteries
- Pens
- A3 paper
- Square post-it notes

A.1.2.1 Provide Participant Information Sheet

Prior to the interview, the participant should be given the 'participant information sheet' and asked to read the positioning text and fill in the demographic information.

A.1.2.2 Verbal Positioning

Before starting the recording, the interviewer should verbally set expectations to the participant by:

- Reiterating the 'positioning' text from the 'participant information sheet'.
- Stating the interviewer's expert knowledge is on data collection, not home-heating systems, so they should not assume any verbal/facial cues relate to the accuracy of their answer.
- Describing the three different sections in brief, and explain that there will be paper-based activities, where, together, the interviewer and participant will 'draw a diagram' to represent how they use, and think about home heating.
- Reassure the participant that the best contribution they can make to the research is to express their own thoughts and experiences with home heating, rather than try to present 'perfect understanding' or 'ideal usage'.
- Emphasize to the participant that there will be opportunities to amend or change their answers throughout the interview, if they feel they haven't described something the way they intended.

A.1.3 What to Expect and How to Deal with It

1. When providing a response, participants may answer questions that belong to different sections. The interviewer should follow the participant's train of thought rather than cut them off. If a question has already been addressed in a different section, the interviewer should state the question, then refer to the answer already given to show they were listening.
2. Participant answers may be contradictory as the interview progresses. This is expected, and the participant should not be challenged on their inconsistencies. People's models or behaviour may vary when presented with different contexts, and this could be a useful research insight.

3. The length of the interview may vary depending on the age and experience of the participant. Older participants may drift off subject and will need to be tactfully returned to the questions. Older/more experienced participants may take longer in the mental model section as they may have a more detailed/complete understanding of the number and role of heating components. Younger/less-experienced participants may take less time as they have a basic/incomplete understanding of the heating system. Allocate longer time periods for older participants. Avoid leading/putting younger participants under pressure to produce more detailed/complete mental model descriptions.

4. Throughout the interview, participants may feel embarrassed when confronted with their realization that they may use the system in a non-optimal way, or may have less understanding of the heating system than they thought. Repeat verbal positioning and reassure participant by reminding them there is no reason that they should have expert knowledge.

5. In the mental model section, more experienced, or older participants, may fall into the role of trying to 'teach' the interviewer how the heating system works and then may get frustrated that the interviewer (who may appear to them as an intelligent adult) is not 'grasping' what they are saying. Remind the participant that to avoid misunderstandings, it is important they describe what they think, even if the interviewer may know what they mean. Sometimes, it helps to ask them to explain what they mean, as if they were talking to a teenager or adult from a hot country, who have not used home heating before.

A.1.4 INTERVIEW OUTPUTS

1. Diagram of a participant's self-reported 'typical weekly schedule' of home-heating use – created with, and validated by, the participant
2. Diagram of the participant's mental model of how the home-heating system functions, showing the components and the relationship between components – created with, and validated by, the participant
3. Audio recording of interview – allowing in-depth analysis of transcripts if required

A.2 QuACk PARTICIPANT INFORMATION SHEET

We are interested in how people think about home heating, and how they use home-heating devices. We think that heating devices can be difficult to understand and/or that it can be difficult to heat your home in the way that you want.

We are not testing you about your knowledge of energy or mechanical systems, but we will ask you questions about how you think things work, to see if this affects the way you use heating devices.

We will ask you about how you use heating for your particular lifestyle. This is to see how well it matches the energy data we have collected, and to understand better your needs and how you use heating devices to meet those needs.

We will also suggest situations to you and then ask you to imagine the effect on heating your home and the way heating devices should be used.

Home heating is a complicated subject. We do not expect anyone to know the most energy-efficient way to run their home. Your answers will help the design of heating and energy-monitoring devices to make it easier for people to be energy-efficient in a way that fits in with their lifestyle.

All your answers will be kept confidential and stored securely.

Please fill in the following information and bring this sheet with you to the interview:

1. Gender (please circle) Male/Female
2. Age Group (please circle) 20–35   36–45   46–55   56–65   66–75 Over 75
3. Occupation..
4. Country of Origin (where you spent most of your childhood).....................
   ....
5. If you have lived outside of the United Kingdom, how long have you lived in countries where it is normal to heat the home? (Approx. no. years)..........
   ....
6. How long have you lived in your current accommodation? (Approx. no. years).....................................
7. What type of accommodation do you currently live in? (Please circle) Flat   Terrace House   Semi-Detached House   Detached House   Other....
   ....
8. How many bedrooms in your current accomodation?.................................
   ....
9. Do you own your current accommodation? (Please circle) Yes/No/Shared Ownership
10. How many people live in your current accommodation? No. Adults (over 18)......................... No. Children (under 18).....................
11. Do you know if your accommodation has insulation? (Please circle) Y/N/Not Sure
12. If Yes/Not sure, please indicate which of the following apply (please circle): Cavity Wall Insulation   Loft Insulation   Double Glazing   Other.................

A.3   QuACk INTERVIEW TEMPLATE FOR HOME-HEATING DOMAIN

A.3.1   BACKGROUND EXPERIENCE IN HOME HEATING

'For this first section, I will be asking about your past experience with home-heating systems so that I get an idea of what may have influenced your ideas about home heating. We will also ask about your attitude to home heating and talk about the names you give to different parts of the home-heating system, so that I can make sure that we are talking about the same thing'.

1. Do you have any specialist knowledge about heating, energy use or thermo-dynamics of buildings?
2. Have you had previous experience with home-heating devices? If so,
   a. What sort of devices were they – can you describe them, or do you know the make?
   b. Approximately how long have you had experience with home-heating devices?
   c. What type of device you are most familiar with? If they struggle – suggest a couple to get them going (e.g. central heating with radiators, electric fires).
3. Which of the following statements best reflects your attitude to energy over the last 3 months? If the participant looks hesitant, verbalize that the study is not aligned to a particular viewpoint, but is interested in how people think and use heating systems and understanding attitudes to heating may help explain this. If the participant cannot choose, ask them to put them in order of importance.
   a. I want to save money
   b. I want to protect the environment
   c. I want to keep warm
   d. Other (e.g. I want to balance cost/comfort)..
4. What home-heating devices do you have in your current accommodation (I will write these down on post-it notes so we can use them later, and so I know what you are referring to in the interview)? Give an example of 'radia-tor' if they seem unsure. For each device, ask them to describe what they look like and what they refer to them as. Agree a terminology that they are comfortable with, but you are clear on the meaning – e.g. 'heating control' for thermostat, 'heating switch' for boiler override, 'big box' for boiler, 'blower' for hot air heater – write BOTH terms on each post-it note to avoid confusion.

Check all questions answered in Section 3.1

A.3.2 BEHAVIOUR

'For this section, I'm going to first be asking about how you use the heating system in your current accommodation over a typical week. I'll use your answers to create a diagram on this template [show template]. Throughout this section, or at any time in the interview, you can make changes to this diagram, if, for example, you remember something you haven't added, or you do not feel it reflects what you do. After this, I'm going to describe some typical home-heating scenarios to you and ask you what you think you would do in each situation'.

1. When asking the questions – replace the terminology in the questions with that agreed with the participants.

2. If some questions have been covered in previous answers, acknowledge the question by verbalizing the answer given previously to show you were paying attention.
    a. When you get a sense of the pattern of use (e.g. which devices, and what settings used over time), start representing this on the template as they talk. Prime the participant verbally, for example 'What I'm going to do now is draw what I think is the typical way that you use home heating in your home over a week, so you can tell me if you think I have the good idea about what you do'.
    b. Draw out a basic idea of what they have said over a week – separating weekday with weekend if different. On the y-axis, choose the appropriate scale for the devices. If there is a combination of devices – you may need different versions of the scale for different devices (e.g. temperature scale, as well as an on/off scale). As you are drawing, talk through what you are doing so the voice recorder can pick up exactly what you mean, and if there is agreement from the participant).
3. Allow the participant to draw on the diagram to show what they mean, if they find this easier.

A.3.2.1 Self-Report on Usage

1. How do you turn the heating on? (Which device? Describe the steps)
2. When do you normally turn the heating on? Make a note of the times of day, and duration to use for the template.
    a. Is there a difference in the way you use heating at the weekend? If so, please describe.
3. Who normally turns the heating on? Is it more than one person, or typically one person?
    a. If more than one person, could you describe when each person normally turns it on (e.g. times of day, how often, how long for, what controls and steps are used)? When annotating different agents on the template, agree on a key with the participant, and label the behaviour patterns to distinguish between them.
4. Do you ever use the thermostat? If so,
    a. Who normally uses the thermostat?
    b. When do you normally turn it up/down?
    c. Why do you normally turn it up/down?
    d. Are there any other situations in which you use the thermostat? If so, what are they? How often does this happen (every day, every week or rarely)?
5. Do you use the thermostats on the radiators? If so,
    a. When do you do this (every day, every week or rarely)?
    b. How do you make adjustments?
    c. Why would you make adjustments?

6. Have you/someone programmed the programmer? If so,
    a. What times and durations is the programmer set to come on and why?
    b. Have you ever bypassed the programmer to turn heating on/off? If so,
        i. What sorts of situations (when and why and how often?)
        ii. What did you do to bypass the programmer?

Paraphrase what you have understood from this section to allow participant to agree/amend.

Check all questions answered in Section 3.2.1

A.3.2.2 Response to Scenarios

'In this section, I'm going to describe some home-heating scenarios to you. These may, or may not, reflect what happens in your own life. If it doesn't reflect your own life, please imagine what you think you would do in this situation. Afterwards, I will ask you how likely this scenario is for your lifestyle. This section often reminds people of ways that they use their home heating that did not come to mind in the previous section. If this happens, we can make adjustments to the home-heating template to better reflect this'.

1. *Scenario 1*: It is winter; you come home to a cold house and want to put the heating on to warm up. What do you do? Let the participant answer freely, then use the following probes if they have not already been answered:
    a. Describe what device you use and how you adjust it?
    b. Why do you use that device, and why do you adjust it in that way?
    c. You want to warm up quickly – would you do anything different?
        i. How likely is this? (every day, every week, rarely...?)
    d. You want the heating to stay on for a long time – would you do anything different?
        i. How likely is this? (every day, every week, rarely...?)
    e. You want the heating to come on straight away – would you do anything different?
        i. How likely is this? (every day, every week, rarely...?)
    f. You want the house to warm up to a specific temperature – would you do anything different?
        i. How likely is this? (every day, every week, rarely...?)
    g. How typical is scenario 1 for your lifestyle during winter? (every day, every week, rarely...?)
2. *Scenario 2*: You have been working on your laptop all morning and are feeling cold from sitting at your desk for too long. What do you do? Let the participant answer freely, then use the following probes if they have not already been answered:
    a. How likely are you to turn on the heating to warm up? (every day, every week, rarely...?)
    b. How typical is scenario 2 for your lifestyle during winter? (every day, every week, rarely...?)

3. *Scenario 3*: The heating is on at the usual time and you have been rushing around (e.g. doing housework, playing with the children, cooking in a warm kitchen or doing exercise). You now feel uncomfortably warm. What do you do? Let the participant answer freely, then use the following probes if they have not already been answered:
   a. How likely are you to turn on the heating to warm up? (every day, every week, rarely…?)
   b. How typical is scenario 3 for your lifestyle during winter? (every day, every week, rarely…?)
4. *Scenario 4*: You are relaxing with your spouse in the evening, the house is pleasantly warm and the heating is on. What do you do?
   a. How likely are you to turn the heating off? (every day, every week, rarely…?)
   b. How typical is this scenario? (every day, every week, rarely…?)

Paraphrase what you have understood from this section to allow participant to agree/amend.

Check all questions answered in Section 3.2.2

Give participant opportunity to make any changes to their typical behaviour diagram (verbalizing adjustments so they can be understood when listening back to the audio recorder).

A.3.3 MENTAL MODEL OF DEVICE FUNCTION

'In this section, I will be asking you how you think the home-heating system in your current home works. We are not interested in knowing the "correct" answer. We are looking to understand what you imagine happens, as this is more likely to affect your behaviour when using heating in the home. Please say what you think, or have a "guess". Afterwards you will be asked how sure you are. Don't worry if things you say do not match things you have said before, it is normal for people to think differently about how things work, when presented with different situations. As you answer the questions, I will write your answers down on the post-it notes and paper, and arrange the post-it notes and draw lines between them. This will help me to build up a picture of what you imagine'.

- Take the A3 plain paper and place the annotated post-it notes next to it.
- When the participant mentions elements in the post-it notes in their response, add this to the paper in an appropriate place.
- If the participant mentions new terms/devices/concepts, add this to a new post-it note and place on the paper in an appropriate place.
- If the participant gives further details or information about a concept, annotate the appropriate post-it note or paper to reflect this.
- For questions 2, 3 and 4, substitute the concepts in brackets, and repeat the question for each relevant post-it note.

1. How can you tell when the heating is on/off?
   a. What do you see, hear, feel or smell?
2. What is the job of this [device]? Use the following prompts to draw out the different elements of the system and the different conduits and dependencies:
   a. What do you think happens when you [adjust] the [device] (e.g. '... turn up/down the thermostat')?
   b. What do you think the [device] is connected to (e.g. thermostat)?
   c. How does the [device] know when to [operate] (e.g. '... the boiler know when to come on/off')?
   d. What happens when you override the [device] (e.g. programmer)?
   After each post-it note has been through this process, paraphrase using the diagram what you think the participant means, and ask them to confirm/amend, before going onto the next post-it note.
3. If you [adjusted] the [device] to its [extreme maximum] setting – can you explain using the diagram, what would imagine happens (e.g. '... turned the thermostat to its maximum temperature')?
4. If you [adjusted] the [device] to its [extreme minimum] setting, can you explain using the diagram, what you imagine would happen (e.g. '... turned the thermostat TO its minimum temperature')?
   After each post-it note has been through this process, paraphrase using the diagram what you think the participant means, and ask them to confirm/amend, before going onto the next post-it note.
5. When you were thinking or describing how the heating system works, can you think of any other devices that work in the same way? Or did any other things come to mind? Use examples participant has offered already of analogies with other devices (e.g. programmer works like an alarm clock, or boiler works like a kettle), but do not suggest analogies yourself. If they hesitate or look uncomfortable, do not pursue this.

'For this last step, I will be asking you to say how confident you are that this diagram reflects how you imagine your home-heating system works. I'm going to go through each part of the diagram, and paraphrase what I think you mean. If you are happy this reflects what you imagine/think (even if you are not sure if this is correct), I will put a 'smiley'. If you are unsure about what you think (i.e. if the diagram reflects what you think makes sense, but you are not sure what you really believe), then I'll put a '?'. If I have misunderstood something, please let me know, so I can amend the diagram to reflect what you imagine'.

- Go through each component and conduit and paraphrase what has been annotated.
- Ask 'does this reflect what you imagine' and wait for a response.
- If they are happy, annotate with a '☺'.

- If they suggest an amendment, make the amendment, then annotate with a '☺'.
- If they are not sure, annotate with a '?'.

Check all questions answered in Section 3.3

'Thank you for participating in this study, the interview is now over'.

Appendix B

APPENDIX

These tables can be used to analyse the output from the QuACk semi-structured interview in Appendix A. It comprises a separate analysis table for categorizing behaviour patterns (part 1) and mental model descriptions (part 2). Walk-through questions are also provided in parts 3 and 4 to guide the analyst in their categorization. The tables are generic and so can be used for any control type, in any domain.

TABLE B.1

Part 1 – Output 1 Analysis Table for Categorizing Behaviour Patterns of Home Heating

1. Control	2. Agents	3. Regularity	4. Frequency	5. Set points	6. Synchronicity	7. Category
• Devices adjusted in a household during a typical week	• Range and type of agents contributing to behaviour patterns	• Of behaviour patterns	• Of adjustments	• How specific • How variable	• Of behaviour patterns with other factors	• Compatible shared theory/ generic theory of device function
Control device	• Single • User (manual)	• Pattern may be completely irregular • Some parts of pattern may be repeated • Changes in regularity may be based on range of lifestyle and system factors	• Frequent adjustments • Exact behaviour pattern not repeated • Minimal periods of no adjustment when	• Value may be specified, approximate or general (turn up/ turn down) • Variable set point value	• Regular and irregular daily activities • Occupancy of dwelling • Type and level of activity carried out by occupants • Changes in additional variables (e.g. external temperature/comfort levels)	Generic Valve (manual)

(Continued)

TABLE B.1 (Continued)
Part 1 – Output 1 Analysis Table for Categorizing Behaviour Patterns of Home Heating

e.g. thermostat				
• Single • User (manual)	• Mainly irregular patterns, but some parts may be repeated (entire daily pattern never repeated) • Changes in regularity based on range of lifestyle, and system factors	• Frequent adjustments of set point when users at home and awake • No adjustments when absent/sleep	• Specific values less important than direction and extent of adjustment (turn up or down, turn right up, or right down) • Considerable variations in set point value based on range of lifestyle and system factors • Set point turned right down at night	• Adjustments coincide with regular activities (e.g. turning right down when going to bed, turning up when getting up, turning down when leaving for work or cooking) • Irregular activities (e.g. turning up when coming home earlier than usual) • Other variables such as external weather (e.g. turning up when it snows outside), or user comfort (e.g. turning down when user hot from exercising)	Valve (Kempton 1986)

(Continued)

TABLE B.1 (*Continued*)

Part 1 – Output 1 Analysis Table for Categorizing Behaviour Patterns of Home Heating

Control device					
• Single/multiple • User (manual)	• Regular/irregular pattern depending on lifestyle *If used as primary control:* • Patterns may be repeated • Occasional variations in patterns • May be infrequent	• Infrequent adjustments • No adjustment when house unoccupied/users asleep • Multiple users may increase frequency if they have different comfort goals *If used as secondary control (e.g. in conjunction with automatic timer):* • Adjustments may only occur occasionally	• Set points values chosen specifically to activate/deactivate heating *If control offers scale:* • Values may vary (e.g. to 'click' or extreme values) *If control offers discrete options:* • Exact values (corresponding to on/off)	• Routine events (if used as primary control) • Non-routine events • Activity types that affect comfort levels (sedentary activities may encourage user switching on heating, high levels of activity encourage switching off)	Generic switch

(*Continued*)

TABLE B.1 (*Continued*)

Part 1 – Output 1 Analysis Table for Categorizing Behaviour Patterns of Home Heating

e.g. thermostat	Regularity based on regularity of lifestyle				Switch (inferred from Peffer 2011)
• Single/ multiple • User (manual)		• Infrequent instances of adjustment • No adjustment when house unoccupied/users asleep • Multiple users may increase frequency if they have different comfort goals	• Set points values chosen specifically to activate/deactivate heating (e.g. to 'click' when boiler can be heard to 'fire up' or extreme increase/decrease in temperature) • Approximate values may be given rather than specific values	Adjustments coincide with regular activities (e.g. going to bed, getting up, leaving for work, cooking) • Irregular activities (e.g. coming home early) • Operators own comfort (too hot/too cold from exercise, sitting still, housework)	

(Continued)

TABLE B.1 (Continued)

Part 1 – Output 1 Analysis Table for Categorizing Behaviour Patterns of Home Heating

Control device					
• Single/ multiple • User(s) (manual only) • User + digital (e.g. manual set points, automated adjustment)	• Regular pattern of adjustment • Changes in pattern may occur according to changes in lifestyle (e.g. at weekend) • If automated agent included, adjustments may occur when occupants asleep/away from the house	• Infrequent adjustments • Pattern repeated periodically (e.g. daily)	• Specific values chosen to maintain heating variable (temperature, time, boiler activity etc.) • Variations in the values chosen match needs of specific regular lifestyle events *Note: If two feedback devices are used together (e.g. home-heating timer and thermostat), the user may keep static set points if the combined automatic adjustments fulfil their lifestyle requirements*	• Regular daily activities • Ad hoc daily activities (if manually controlled, or if easy to make adjustments to automatic controls)	Generic feedback – automatic on/off

(Continued)

TABLE B.1 (*Continued*)

Part 1 – Output 1 Analysis Table for Categorizing Behaviour Patterns of Home Heating

| e.g. thermostat | • Single
• User (manual) | • Regular pattern of adjustment based around regularity of lifestyle (e.g. set to 20°C on rising, turned to 18°C when leaving the house, turned back to 20°C on returning)
• Intervals between adjustments may vary based on lifestyle, but not comfort levels | • Infrequent adjustments (based on frequency of regular activities)
• Approx. pattern of adjustment repeated daily | • Predetermined set value for when house is occupied, when unoccupied
• Night set point chosen, but not significantly different from time set point | • Adjustments coincide with regular activities (e.g. getting up, leaving for work, coming home, going to bed)
• Ad hoc adjustments rarely made–and only based on changes to activity/occupancy, *not* comfort | Feedback (Kempton 1986) |

(Continued)

TABLE B.1 (*Continued*)

Part 1 – Output 1 Analysis Table for Categorizing Behaviour Patterns of Home Heating

Control device					
• Single/duel • User (manual) • User (manual)+ digital (automatic)	• Regularity of pattern of adjustment, based on regularity of lifestyle	• Frequency of adjustments depends on the set point chosen by the user (greater set point, greater interval between adjustments, lower frequency) • Frequency of pattern of adjustment, dependent on lifestyle/comfort levels	• Specific set point values chosen to determine 'automatic off' of heating variable • Set point value may vary depending on lifestyle/comfort • May be a 'reset' set point value	• Regular daily activities • Irregular activities • Comfort levels	Generic feedback – automatic off

(Continued)

TABLE B.1 (*Continued*)

Part 1 – Output 1 Analysis Table for Categorizing Behaviour Patterns of Home Heating

e.g. thermostat				
• Single • User (manual on)	• Regularity of pattern of adjustments depends on regularity of lifestyle	• Frequency of adjustments depends on set point chosen. If high temperature value chosen, not likely to make further adjustments for a while • Frequency of pattern of adjustment dependent on lifestyle/comfort levels	• Set point values chosen to determine how long heating on – higher set point to ensure heating is on for a longer time • Lower set point to ensure heating is on for shorter period of time	• Adjustments coincide with regular activities where heat levels are too low (e.g. getting up, returning from work) • Irregular activities (e.g. coming home early to cold house) • Operators own comfort (too cold from sitting still)	Timer (inferred from Norman 1986)

TABLE B.2
Part 2 – Output 2 Analysis Table for Categorizing Mental Model Descriptions of Home Heating

1. Control	2. Input behaviour	3. Key variable	4. Key element	5. Sensor	6. Sensed variable	7. Rule	8. Category
• Operated by user to adjust home heating	• Afforded by control device for specific user • Relates to key/sensed variable • May initiate automatic adjustments	• Related to energy consumption (dependent on research question)	• Influences key variable • Linked to control device/sensor (directly or indirectly)	• Measures sensed variable • Linked to Key element (directly/indirectly)	• Measured by sensor • Value used to enable automatic adjustment	• How variations in input behaviour affects the key variable • Criteria for automatic adjustments of key variable • Including the role of a sensor, sensed variable and key element (where relevant)	• Generic/shared theory(manual/automatic)
Control	*Control* • Allows continuous/interval adjustments that affect the key variable	*Key variable* • Controlled by key element	*Key element* • May house control	None	n/a	*Variations in input behaviour of control are linearly related to: variations in key variable*	Generic valve (manual)

(Continued)

TABLE B.2 (Continued)

Part 2 – Output 2 Analysis Table for Categorizing Mental Model Descriptions of Home Heating

e.g. thermostat knob	*Thermostat knob* • Allows increase/decrease temperature dial to adjust *boiler intensity*	Boiler intensity	Thermostat	None	n/a	*Increasing/decreasing* the *thermostat knob* results in increases/decreases in *boiler intensity*	Valve (Kempton 1986)
Control	*Control* • Allows discrete modes (e.g. on/off) that affect the *key variable*	*Key variable* • Controlled by *key element*	*Key element* • May house *control*	None	n/a	When *control* is *activated*, it *enables key variable* When *control* is *deactivated*, it *disables key variable*	Generic switch (manual)
e.g. thermostat knob	*Thermostat knob* • Allows increase/decrease temperature. dial to turn on/off *boiler operation*	Boiler operation	Thermostat	None	n/a	When *thermostat knob* is turned up, it activates *boiler operation*. When the *thermostat knob* is turned down, it deactivates *boiler operation*	Switch (Peffer 2011)

(Continued)

TABLE B.2 (Continued)
Part 2 – Output 2 Analysis Table for Categorizing Mental Model Descriptions of Home Heating

Control device	Control device	Key variable	Key element	Sensor	Sensed variable	Control	Generic feedback (automatic on/off)
	• Determines/initiates target value of sensed variable • Influences key variable (or other variables) • Enables automatic maintenance of target value of sensed variable	• Controlled by key element • Changes result in changes to sensed variable	• Influences key variable • Receives input from sensor (directly/indirectly)	• detects sensed variable and feeds back to key element (directly/indirectly)	• Measured by sensor	• Determines/initiates target value of sensed variable • This value is compared to current value of sensed variable, measured by the sensor • If target value is higher than current value, the key element will enable the key variable • If lower/same, it will disable the key variable	

(Continued)

TABLE B.2 (Continued)

Part 2 – Output 2 Analysis Table for Categorizing Mental Model Descriptions of Home Heating

e.g.	Thermostat knob	Boiler operation	Thermostat	Thermometer	Room temperature	Feedback (Kempton 1986)
Thermostat knob	*Thermostat knob* • Lets me set my desired *room temperature* • It influences when the boiler comes on and off • This room temperature is then automatically maintained					*Thermostat knob* • Determines the target room temperature • *Target room temperature* is compared to *current room temperature* measured by *thermometer* • If *target room temperature* is higher, the *thermostat* enables *boiler operation*. If lower/same, it disables *boiler operation*

(Continued)

TABLE B.2 (Continued)

Part 2 – Output 2 Analysis Table for Categorizing Mental Model Descriptions of Home Heating

Control	*Control*	*Key variable*	*Key element*	*Sensor*	*Sensed variable*	*Control*	Generic feedback
	• Activates *key variable* • Determines/initiates *target* value of *sensed variable* • Automatically disables *key variable* when target value reached	• Controlled by *key element* • Changes result in changes to *sensed variable*	• Influences *key variable* • Receives input from *sensor* (directly/indirectly)	• Detects *sensed variable* and feeds back to *key element* (directly/indirectly)	• Measured by *sensor*	• Activates *key variable* AND determines/initiates *target* value of *sensed variable* • This value is compared to *current* value of *sensed variable*, measured by the *sensor* • When target value is reached, the *key element disables the key variable*	(manual on/ automatic off)

(Continued)

TABLE B.2 (Continued)

Part 2 – Output 2 Analysis Table for Categorizing Mental Model Descriptions of Home Heating

	Boiler operation	Thermostat	Timer	Time period of boiler operation	Timer (Norman 2002)
e.g. thermostat knob	*Thermostat knob* • Activates *boiler operation* • Lets me set my desired *time period of boiler operation* • Automatically disables *boiler operation* when this time period is reached			Time period of boiler operation	*Thermostat knob* • Activates *boiler operation* AND determines *target* value of *time period of boiler operation* • This value is compared to *current* value of *time period of boiler operation* measured by the *timer* • When time period is reached, the *thermostat disables boiler operation*

TABLE B.3

Part 3 – Walk-Through Questions to Guide Analysts When Categorizing Output 1 from QuACk

Q1. Control	Q2. Agents	Q3. Regularity	Q4. Frequency	Q5. Set points	Q6. Synchronicity	Q7. Association
What heating controls does the user adjust in a typical week?	a. How many agents (human or digital) are responsible for creating the pattern depicted? b. Who are the agents? c. When do different agents come into play?	a. How regular is the pattern of adjustments? b. When do changes in regularity occur?	a. How frequent are the adjustments b. When do changes in frequency occur?	a. How specific are the set points values? b. How do set point values vary?	What do the variations in the pattern coincide with?	a. Can the intention of the user be identified in the behaviour pattern? b. Has the user represented this device in their mental model description?
e.g. • Thermostat • Programmer • TRV • Boiler override • Boiler water temp • Manual on/ off switch etc.	e.g. a. • Single • Dual • Multiple b. • User only • User and housemate • Housemate only • User and automatic agent • Housemate and automatic agent	e.g. a. • Regular daily pattern of adjustment • Irregular evening pattern of adjustment • Regular daytime pattern of adjustment, etc.	e.g. a. • Frequent • Infrequent b. • Frequent during the evening and infrequent otherwise • Infrequent during the week and frequent during the weekend	e.g. a. • Specific values described • Approximate values described • General increases/ decreases shown	e.g. • Routine events (school times, work times, bedtimes, waking up times) • Non-routine events (home early, leaving late, day-trips, household party)	e.g. a. • Yes (single agent, multiple agents but intention of different agents is clear, multiple agents fulfilling joint intention) • No (multiple agents confusing intention of user)

(Continued)

TABLE B.3 (Continued)

Part 3 – Walk-Through Questions to Guide Analysts When Categorizing Output 1 from QuACk

c.
- User adjusts set values, automatic agent makes adjustments according to set values
- User makes adjustments during week, housemate at the weekend
- User makes adjustments during day, housemate in the evening,

etc.

b.
- Regular pattern during week, but irregular at weekend
- Regular pattern during the day, but irregular pattern in evening
- Regular pattern all the time
- Irregular pattern of use, but regular intervals when adjusted

etc.

b.
- Always frequent
- Always infrequent

etc.

Note: *The analyst needs to infer what constitutes 'frequent' or 'infrequent', depending on the type of control*

b.
- Large variations in set point values
- Static set point values
- Minor variations in set point values
- Static set point values during the week, variations at the weekend
- Static set point values during the day, large variations in the evening

etc.

- Changes in other variables (comfort levels, external temperatures)
- Changes in activity levels (inactive, active)
- Changes in activity types (cooking, exercising, sleeping, studying, watching TV)

b.
- Yes
- No – control device absent from output 2

TABLE B.4

Part 4 – Walk-Through Questions to Guide Analysts When Categorizing Output 2 from QuACk

Q1. *Control device*	Q2. *Input behaviour*	Q3. *Key variable*	Q4. *Key element*	Q5. *Sensor*	Q6. *Sensed variable*	Q7. *Rule*
What specific element does the user *directly* interact with?	a. What adjustment does this specific user do with the *control device?* b. What does the user believe they are influencing? c. Does the user describe/ imply the system making 'automatic' adjustments?	What energy-consuming variable is the user trying to influence when they adjust the *control device?*	Which element does the user believe is responsible for controlling the key *variable?*	Does the user describe/imply a 'sensing' element that measures a variable to enable 'automatic' adjustments? If not, go to Q7	What variable does the user describe/ imply as being measured by the sensor?	What rule can be constructed from the user's mental model to describe: how the *input behaviour* affects the *key variable,* including the role of a *sensor, sensed variable* and *key element* (*where appropriate*)? (*Continued*)

TABLE B.4 (Continued)

Part 4 – Walk-Through Questions to Guide Analysts When Categorizing Output 2 from QuACk

e.g.	e.g.	e.g.	e.g.	e.g.	e.g.	
• Thermostat dial • Boiler on/off switch • Programmer schedule • Programmer override • TRV knob • Boiler temperature control	a. Setting target or specific value of variable • Changing existing value of variable on a scale • Selecting on/off, etc. b. Temperature (house, room, water, body) • Intensity (boiler, water/gas flow) • Duration (boiler activation, time periods for activation, etc.) c. 'It' turns on/off according to... • 'It' turns itself off... • 'It' maintains a target value of...etc.	• Length of boiler activation periods • Length of boiler activation • Intensity of boiler • Amount of heat transferred to radiators • Amount of heat emitted from radiators • Amount of water heated • Radiator temperature • Temperature of water • Speed of water flow	• Boiler • Programmer • Thermostat • TRV	• Clock • Thermometer • Timer • Heat sensor • Flow sensor • Body temperature sensor	• House temperature • Room temperature • Water/gas temperature • Boiler intensity • Water flow rate • Body temperature • Length of time boiler has been operating • Length of time	Compare with examples of generic and shared theories of home heating.

References

Aerts, D., Minnen, J., Glorieux, I., Wouters, I. and Descamps, F., 2014. A method for the identification and modelling of realistic domestic occupancy sequences for building energy demand simulations and peer comparison. *Building and Environment*, 75, 67–78.

Alwitt, L. F. and Pitts, R. E., 1996. Predicting purchase intentions for an environmentally sensitive product. *Journal of Consumer Psychology*, 5 (1), 49–64.

Anderson, J. R., 1983. *The Architecture of Cognition*. Cambridge, MA: Harvard University Press.

Annett, J., 2002. A note on the validity and reliability of ergonomics methods. *Theoretical Issues in Ergonomics Science*, 3 (2), 228–232.

Bainbridge, L., 1992. Mental models in cognitive skill: The example of industrial process operation. In: Rogers, Y., Rutherford, A. and Bibby, P. A. (eds.), *Models in the Mind: Theory, Perspective and Applications*. London, UK: Academic Press, pp. 119–143.

Ball, L. J. and Christensen, B. T., 2009. Analogical reasoning and mental simulation in design: Two strategies linked to uncertainty resolution. *Design Studies*, 30 (2), 169–186.

Bartlett, F. C., 1932. *Remembering: A Study in Experimental and Social Psychology*. Cambridge, England: Cambridge University Press.

Baxter, G., Besnard, D. and Riley, D., 2007. Cognitive mismatches in the cockpit: Will they ever be a thing of the past? *Applied Ergonomics*, 38 (4), 417–423.

BBC, 2014. Smart meters will save only 2% on energy bills, say MPs [Online]. Available: http://www.bbc.co.uk/news/business-29125809 [Accessed 25 November 2014].

Bourbousson, J., Poizat, G., Saury, J. and Seve, C., 2011. Description of dynamic shared knowledge: An exploratory study during a competitive team sports interaction. *Ergonomics*, 54 (2), 120–138.

Bowers, C. A. and Jentsch, F., 2005. Team workload. In: Stanton, N. A., Hedge, A., Brookhuis, K., Salas, E. and Hendricks, H. (eds.), *Handbook of Human Factors and Ergonomic Methods*. London, UK: Taylor & Francis Group, pp. 57–61.

Branaghan, R. J., Covas-Smith, C. M., Jackson, K. D. and Eidman, C., 2011. Using knowledge structures to redesign an instructor-operator station. *Applied Ergonomics*, 42 (6), 934–940.

Brewer, W. F., 1987. Schemas versus mental models in human memory. In: Morris, P. (ed.), *Modelling Cognition*, Chichester, U.K.: Wiley, pp. 187–197.

Brown, Z. and Cole, R. J., 2009. Influence of occupants' knowledge on comfort expectations and behaviour. *Building Research and Information*, 37 (3), 227–245.

Buxton, W., 1986. There's more to interaction than meets the eye: Some issues in manual input. In: Norman, D. A. and Draper, S. W. (eds.), *User Centered System Design: New Perspectives on Human–Computer Interaction*. Hillsdale, NJ: Lawrence Erlbaum Associates, Inc.

Carroll, J. M. and Olson, J. R. (eds.), 1987. *Mental Models in Human–Computer Interaction: Research Issues about What the User of Software Knows*. Washington, DC: National Academy Press.

Chetty, M., Tran, D. and Grinter, R. E., 2008. Getting to green: Understanding resource consumption in the home. *UbiComp'08: Proceedings of the 10th International Conference on Ubiquitous Computing*. Seoul, South Korea: ACM.

Climate Change Act, 2008. [Online]. London, UK: Department of Energy & Climate Change. Available: http://www.decc.gov.uk/en/content/cms/legislation/cc_act_08/cc_act_08.aspx [Accessed 10 November 2011].

Collins, A. and Gentner, D., 1987. How people construct mental models. In: Holland, D. and Quinn, N. (eds.), *Cultural Models in Language and Thought*. Cambridge, UK: Cambridge University Press, pp. 243–265.

Combe, N., Harrison, D., Dong, H., Craig, S. and Gill, Z., 2011. Assessing the number of users who are excluded by domestic heating controls. *International Journal of Sustainable Engineering*, 4 (1), 84–92.

Connell, I. W., 1998. Error analysis of ticket vending machines: Comparing analytic and empirical data. *Ergonomics*, 41 (7), 927–961.

Craik, K. J. W., 1943. *The Nature of Explanation*. Cambridge, England: Cambridge University Press.

Crossman, E. R. F. W. and Cooke, J. E., 1974. Manual control of slow-response systems. In: Edwards, E. and Lees, F. (eds.), *The Human Operator in Process Control*. London, UK: Taylor & Francis, pp. 51–66.

Cuomo, D. and Bowen, C., 1994. Understanding usability issues addressed by three user-system interface evaluation techniques. *Interacting with Computers*, 6 (1), 86–108.

Cutica, I. and Bucciarelli, M., 2011. "The more you gesture, the less i gesture": Co-speech gestures as a measure of mental model quality. *Journal of Nonverbal Behavior*, 35 (3), 173–187.

Dalla Rosa, A. and Christensen, J. E., 2011. Low-energy district heating in energy-efficient building areas. *Energy*, 36 (12), 6890–6899.

Darby, S., 2001. Making it obvious: Designing feedback into energy consumption. In: Bertoldi, P., Ricci, A. and Dealmeida, A. (eds.), *Second International Conference on Energy Efficiency in Household Appliances and Lighting*. Naples, Italy: Springer.

DECC, 2013. Smarter heating controls research programme [Online]. London, UK: Department of Energy & Climate Change. Available: https://www.gov.uk/government/policies/helping-households-to-cut-their-energy-bills/supporting-pages/smarter-heating-controls-research-programme [Accessed 9 February 2016].

De Kleer, J. and Brown, J. S., 1983. Assumptions and ambiguities in mechanistic mental models. In: Gentner, D. and Stevens, A. L. (eds.), *Mental Models*. Hillsdale, NJ: Lawrence Erlbaum Associates, pp. 155–190.

Desnoyers, L., 2004. The role of qualitative methodology in ergonomics: A commentary. *Theoretical Issues in Ergonomics Science*, 5 (6), 495–498.

Edwards, J., 2005. Subtext: Uncovering the simplicity of programming. *OOPSLA'05: Proceedings of the 20th Annual ACM SIGPLAN Conference on Object Oriented Programming, Systems, Languages, and Applications*. New York: ACM Press, pp. 505–518.

Emery, A. F. and Kippenhan, C. J., 2006. A long term study of residential home heating consumption and the effect of occupant behavior on homes in the Pacific Northwest constructed according to improved thermal standards. *Energy*, 31 (5), 677–693.

Energy Saving Trust, 2013. Thermostats and controls [Online]. Available: https://www.gov.uk/government/uploads/system/uploads/attachment_data/file/228752/9780108508394.pdf [Accessed 17 April 2013].

Evans, J. S. T., Clibbens, J. and Rood, B., 1995. Bias in conditional inference – Implications for mental models and mental logic. *Quarterly Journal of Experimental Psychology Section a-Human Experimental Psychology*, 48 (3), 644–670.

Fabi, V., Andersen, R. V., Corgnati, S. and Olesen, B. W., 2012. Occupants' window opening behaviour: A literature review of factors influencing occupant behaviour and models. *Building and Environment*, 58, 188–198.

Field, A., 2000. *Discovering Statistics using SPSS for Windows*, Vol. 2, London: Sage Publications, pp. 44–322.

Fischer, C., 2008. Feedback on household electricity consumption: A tool for saving energy? *Energy Efficiency*, 1 (1), 79–104.

Flyvbjerg, B., 2011. Case study. In: Denzin, N. K. and Lincoln, Y. S. (eds.), *The Sage Handbook of Qualitative Research*, 4th edn. Thousand Oaks, CA: Sage, pp. 301–316.

Frède, V., Nobes, G., Frappart, S., Panagiotaki, G., Troadec, B. and Martin, A., 2011. The acquisition of scientific knowledge: The influence of methods of questioning and analysis on the interpretation of children's conceptions of the earth. *Infant and Child Development*, 20 (6), 432–448.

Gentner, D. and Gentner, D. R., 1983. Flowing waters or teeming crowds: Mental models of electricity. In: Gentner, D. and Stevens, A. L. (eds.), *Mental Models*. Hillsdale, NJ: Lawrence Erlbaum Associates, pp. 99–129.

Glad, W., 2012. Housing renovation and energy systems: The need for social learning. *Building Research and Information*, 40 (3), 274–289.

Gram-Hanssen, K., 2010. Residential heat comfort practices: Understanding users. *Building Research & Information*, 38 (2), 175–186.

Greene, J. A. and Azevedo, R., 2007. Adolescents' use of self-regulatory processes and their relation to qualitative mental model shifts while using hypermedia *Journal of Educational Computing Research*, 36 (2), 125–148.

Grote, G., Kolbe, M., Zala-Mezö, E., Bienefeld-Seall, N. and Künzle, B., 2010. Adaptive coordination and heedfulness make better cockpit crews. *Ergonomics*, 53 (2), 211–228.

Guerra-Santin, O. and Itarda, L., 2010. Occupants' behaviour: Determinants and effects on residential heating consumption. *Building Research & Information*, 38 (3), 318–338.

Gupta, M., Intille, S. S., and Larson, K., May 2009. Adding gps-control to traditional thermostats: An exploration of potential energy savings and design challenges. In: *International Conference on Pervasive Computing*. Berlin, Heidelberg: Springer, pp. 95–114.

Halasz, F. G. and Moran, T. P., December 1983. Mental models and problem solving in using a calculator. In: *Proceedings of the SIGCHI conference on Human Factors in Computing Systems*. New York, ACM, pp. 212–216.

Hancock, P. A., Hancock, G. M. and Warm, J. S., 2009. Individuation: The N = 1 revolution. *Theoretical Issues in Ergonomics Science*, 10 (5), 481–488.

Hancock, P. A. and Szalma, J. L., 2004. On the relevance of qualitative methods for ergonomics. *Theoretical Issues in Ergonomic Science*, 5 (6), 499–506.

Hanisch, K. A., Kramer, A. F. and Hulin, C. L., 1991. Cognitive representations, control, and understanding of complex systems: A field study focusing on components of users' mental models and expert/novice differences. *Ergonomics*, 34 (8), 1129–1145.

Hignett, S. and Wilson, J. R., 2004. The role for qualitative methodology in ergonomics: A case study to explore theoretical issues. *Theoretical Issues in Ergonomics Science*, 5 (6), 473–493.

Hutchins, E., 1983. Understanding Micronesian navigation. In: Gentner, D. and Stevens, A. L. (eds.), *Mental Models*. Hillsdale, NJ: Lawrence Erlbaum, pp. 191–225.

Hutchins, E., Hollan, J. and Norman, D., 1985. Direct manipulation interfaces. *Human Computer Interaction*, 1 (4), 311–338.

Ifenthaler, D., Masduki, I. and Seel, N. M., 2011. The mystery of cognitive structure and how we can detect it: Tracking the development of cognitive structures over time. *Instructional Science*, 39 (1), 41–61.

Jagacinski, R. J. and Miller, R. A., 1978. Describing the human operator's internal model of a dynamic system. *Human Factors*, 20, 425–433.

Jenkins, D. P., Salmon, P. M., Stanton, N. A. and Walker, G. H., 2010. A new approach for designing cognitive artefacts to support disaster management. *Ergonomics*, 53 (5), 617–635.

Jenkins, D. P., Salmon, P. M., Stanton, N. A., Walker, G. H. and Rafferty, L., 2011. What could they have been thinking? How sociotechnical system design influences cognition: A case study of the Stockwell shooting. *Ergonomics*, 54 (2), 103–119.

Johnson-Laird, P. N., 1983. *Mental Models: Towards a Cognitive Science of Language, Inference and Consciousness*. Cambridge, UK: Cambridge University Press.

Johnson-Laird, P. N., 1986. Reasoning without logic. *Reasoning and Discourse Processes*, 13–49.

Johnson-Laird, P. N., 1987. The mental representation of the meaning of words. *Cognition*, 25 (1), 189–211.

Johnson-Laird, P. N., 1989. Mental models. In: Posner, M. I. (ed.), *Foundations of Cognitive Science*. Cambridge, MA: MIT Press, pp. 469–499.

Johnson-Laird, P. N., 2005. *The History of Mental Models [Online]*. Princeton, NJ: Princeton University. Available: http://mentalmodels.princeton.edu/publications/ [Accessed 4 December 2010].

Jones, P. E. and Roelofsma, P. H. M. P., 2000. The potential for social contextual and group biases in team decision-making: biases, conditions and psychological mechanisms. *Ergonomics*, 43 (8), 1129–1152.

Kahneman, D. and Tversky, A., 1982. The simulation heuristic. In: Kahneman, D., Slovic, P. and Tversky, A. (eds.), *Judgement under Uncertainty: Heuristics and Biases*. Cambridge, UK: Cambridge University Press, pp. 201–208.

Kaiser, F. G., Wolfing, S. and Fuhrer, U., 1999. Environmental attitude and ecological behaviour. *Journal of Environmental Psychology*, 19 (1), 1–19.

Kalantzis, A., Thatcher, A. and Sheridan, C., 2016. Mental models of a water management system in a green building. *Applied ergonomics*, 57, 36–47.

Kanis, H., 2002. Can design supportive research be scientific? *Ergonomics*, 45 (14), 1037–1041.

Kanis, H., 2004. The quantitative-qualitative research dichotomy revisited. *Theoretical Issues in Ergonomics Science*, 5 (6), 507–516.

Kaplan, K., 2009. Memo to stakeholders on suspension of programmable thermostat specification [Online]. Washington, DC: U.S. Environmental Protection Agency. Available: https://www.energystar.gov/index.cfm?c=archives.thermostats_spec [Accessed 25 November 2014].

Kempton, W., 1986. Two theories of home heat control. *Cognitive Science*, 10, 75–90.

Kempton, W., 1987. Two theories of home heat control. In: Holland, D. and Quinn, N. (eds.), *Cultural Models in Language and Thought*. Cambridge, UK: Cambridge University Press, pp. 222–241.

Kennedy, D. M. and Mccomb, S. A., 2010. Merging internal and external processes: Examining the mental model convergence process through team communication. *Theoretical Issues in Ergonomics Science*, 11 (4), 340–358.

Kessel, C. J. and Wickens, C. D., 1982. The transfer of failure-detection skills between monitoring and controlling dynamic systems. *Human Factors: The Journal of the Human Factors and Ergonomics Society*, 24, 49–60.

Kieras, D., Meyer, D. and Ballas, J., 2001. Towards demystification of direct manipulation: Cognitive modeling charts the gulf of execution. *Proceedings of CHI 2001*. New York: ACM, pp. 128–135.

Kieras, D. E. and Bovair, S., 1984. The role of a mental model in learning to operate a device. *Cognitive Science*, 8 (3), 255–273.

Klauer, K. C. and Musch, J., 2005. Accounting for belief bias in a mental model framework? No problem! Reply to Garnham and Oakhill (2005). *Psychological Review*, 112 (2), 519–520.

Klauer, K. C., Musch, J. and Naumer, B., 2000. On belief bias in syllogistic reasoning. *Psychological Review*, 107, 852–884.

Ko, A. J., Myers, B. A. and Aung, H. H., 2004. Six learning barriers in end-user programming systems. *Proceedings of the 2004 IEEE Symposium on Visual Languages and Human Centric Computing*, Rome, Italy.

Kuo-Ming, C., Shah, N., Farmer, R. and Matei, A., 2012. Energy management system for domestic electrical appliances. *International Journal of Applied Logistics*, 3 (4), 48–60.

Lakoff, G. and Johnson, M., 1981. *Metaphors We Live By*. Chicago, IL: University of Chicago Press.

Langan-Fox, J., Wirth, A., Code, S., Langfield-Smith, K. and Wirth, A., 2001. Analyzing shared and team mental models. *International Journal of Industrial Ergonomics*, 28 (2), 99–112.

Larsson, A. F., 2012. Driver usage and understanding of adaptive cruise control. *Applied Ergonomics*, 43 (3), 501–506.

Lenior, D., Janssen, W., Neerincx, M. and Schreibers, K., 2006. Human-factors engineering for smart transport: Design support for car drivers and train traffic controllers. *Applied Ergonomics*, 37 (4), 479–490.

Leung, C. and Ge, H., 2013. Sleep thermal comfort and the energy saving potential due to reduced indoor operative temperature during sleep. *Building and Environment*, 59, 91–98.

Lienhard, J. H., 2011. *A Heat Transfer Textbook: Dover Civil and Mechanical Engineering*. Mineola, NY: Dover Publications.

Lilley, D., 2009. Design for sustainable behaviour: Strategies and perceptions. *Design Studies*, 30 (6), 704–720.

Lockton, D., Harrison, D. and Stanton, N. A., 2010. The design with intent method: A design tool for influencing user behaviour. *Applied Ergonomics*, 41 (3), 382–392.

Lutzenhiser, L., 1993. Social and behavioral aspects of energy use. *Annual Review of Energy and the Environment*, 18, 247–289.

Lutzenhiser, L. and Bender, S., 2008. The average American unmasked: Social structure and difference in household energy use and carbon emissions. *ACEEE Summer Study on Energy Efficiency in Buildings*. Washington, DC: ACEEE.

Mack, Z. and Sharples, S., 2009. The importance of usability in product choice: A mobile phone case study. *Ergonomics*, 52 (12), 1514–1528.

Mahapatraa, K., Naira, G. and Gustavssona, L., 2011. Swedish energy advisers' perceptions regarding and suggestions for fulfilling homeowner expectations. *Energy Policy*, 39 (7), 4264–4273.

Manktelow, K. and Jones, J., 1987. Principles from the psychology of thinking and mental models. In: Gardiner, M. M. and Christie, B. (eds.), *Applying Cognitive Psychology to User-Interface Design*. Chichester, UK: Wiley, pp. 83–117.

Mathieu, J. E., Heffner, T. S., Goodwin, G. F., Salas, E. and Cannon-Bowers, J. A., 2000. The influence of shared mental models on team process and performance. *Journal of Applied Psychology*, 85 (2), 273–283.

Mccloskey, M., 1983. Naive theories of motion. In: Gentner, D. and Stevens, A. L. (eds.), *Mental Models*. Hillsdale, NJ: Lawrence Erlbaum Associates, pp. 299–324.

Meier, A. K., Aragon, C., Peffer, T., Perry, D. and Pritoni, M., 2011. Usability of residential thermostats: Preliminary investigations. *Building and Environment*, 46, 1891–1898.

Meister, D., 1977. Human error in man-machine systems. In: Brown, S. C. and Martin, J. N. T. (eds.), *Human Aspects of Man-Made Systems*. Milton Keynes, UK: Open University Press.

Mohageg, M. F., 1991. Object-oriented versus bit-mapped graphics interfaces: Performance and preference differences for typical applications. *Behaviour & Information Technology*, 10 (2), 121–147.

Moray, N., 1990. Designing for transportation safety in the light of perception, attention, and mental models. *Ergonomics*, 33 (10), 1201–1213.

Newstead, S. E. and Evans, J. S. T., 1993. Mental models as an explanation of belief bias effects in syllogistic reasoning. *Cognition*, 46 (1), 93–97.

Newstead, S. E., Pollard, P., Evans, J. S. B. T. and Allen, J. L., 1992. The source of belief bias effects in syllogistic reasoning. *Cognition*, 45 (3), 257–284.

Norman, D. A., 1983. Some observations on mental models. In: Gentner, D. and Stevens, A. L. (eds.), *Mental Models*. Hillsdale, NJ: Lawrence Erlbaum Associates, pp. 7–14.

Norman, D. A., 1986. Cognitive engineering. In: Norman, D. A. and Draper, S. W. (eds.), *User Centered System Design: New Perspectives on Human–Computer Interaction*. Hillsdale, NJ: Lawrence Erlbaum Associates, pp. 31–61.

Norman, D. A., 1993. *Things that Make Us Smart: Defending Human Attributes in the Age of the Machine*. Hillsdale, NJ. Basic Books.

Norman, D. A., 2002. *The Design of Everyday Things*. New York: Basic Books.

Oakhill, J. V. and Johnson-Laird, P. N., 1985. The effects of belief on the spontaneous production of syllogistic conclusions. *The Quarterly Journal of Experimental Psychology Section A*, 37 (4), 553–569.

Oakhill, J., Johnson-Laird, P. N. and Garnham, A., 1989. Believability and syllogistic reasoning. *Cognition*, 31 (2), 117–140.

Oppenheim, A. N., 2000. *Questionnaire Design, Interviewing and Attitude Measurement*. Bloomsbury Publishing, Pinter Publications, New York.

Ormerod, T. C., Manktelow, K. I. and Jones, G. V., 1993. Reasoning with three types of conditional: Biases and mental models. *Quarterly Journal of Experimental Psychology*, 46A (4), 653–677.

Papakostopoulos, V. and Marmaras, N., 2012. Conventional vehicle display panels: The drivers' operative images and directions for their redesign. *Applied Ergonomics*, 43 (5), 821–888.

Payne, S. J., 1991. A descriptive study of mental models. *Behaviour & Information Technology*, 10 (1), 3–21.

Peffer, T., Pritoni, M., Meier, A., Aragon, C., and Perry, D., 2011. How people use thermostats in homes: A review. *Building and Environment*, 46 (12), 2529–2541.

Peffer, T., Daniel, P., Marco, P., Cecilia, A. and Alan, M., 2013. Facilitating energy savings with programmable thermostats: Evaluation and guidelines for the thermostat user interface. *Ergonomics*, 56 (3), 463–479.

Pierce, J., Schiano, D. J., Paulos, E. and ACM, 2010. *Home, Habits, and Energy: Examining Domestic Interactions and Energy Consumption*. New York: Association for Computing Machinery.

Plant, K. L. and Stanton, N. A., 2016. The development of the Schema World Action Research Method (SWARM) for the elicitation of perceptual cycle data. *Theoretical Issues in Ergonomics Science*, 17 (4), 376–401.

Quayle, J. D. and Ball, L. J., 2000. Working memory, metacognitive uncertainty, and belief bias in syllogistic reasoning. *The Quarterly Journal of Experimental Psychology Section A*, 53 (4), 1202–1223.

Raaij, W. F. V. and Verhallen, T. M. M., 1983. Patterns of residential energy behavior. *Journal of Economic Psychology*, 4, 85–106.

Rafferty, L. A., Stanton, N. A. and Walker, G. H., 2010. The famous five factors in teamwork: A case study of fratricide. *Ergonomics*, 53 (10), 1187–1204.

Rasmussen, J., 1983. Skill, rules and knowledge: Signals, signs, and symbols, and other distinctions in human performance models. *IEEE Transactions on Systems, Man and Cybernetics*, 13 (3), 257–266.

Rasmussen, J. and Jensen, A., 1974. Mental procedures in real-life tasks: A case study of electronic trouble shooting. *Ergonomics*, 17 (3), 293–307.

Rasmussen, J. and Rouse, W. B. (eds.), 1981. *Human Detection and Diagnosis of System Failures*. New York: Plenum Press.

Reason, J., 1990. *Human Error*. Cambridge, UK: Cambridge University Press.

Reber, A. S., 1985. *The Penguin Dictionary of Psychology*. London, England: Penguin Group.

Revell, K., 2014. Estimating the environmental impact of home energy visits and extent of behaviour change. *Energy Policy*, 73, 461–470.

Revell, K. M. A. and Stanton, N. A., 2012. Models of models: Filtering and bias rings in depiction of knowledge structures and their implications for design. *Ergonomics*, 55 (9), 1073–1092.

Revell, K. M. A. and Stanton, N. A., 2014. Case studies of mental models in home heat control: Searching for feedback, valve, timer and switch theories. *Applied Ergonomics*, 45 (3), 363–378.

Revell, K. M. and Stanton, N. A., 2015. When energy saving advice leads to more, rather than less, consumption. *International Journal of Sustainable Energy*, 36 (1), 1–19.

Revell, K. M. and Stanton, N. A., 2016. Mind the gap–Deriving a compatible user mental model of the home heating system to encourage sustainable behaviour. *Applied Ergonomics*, 57, 48–61.

Richardson, M. and Ball, L. J., 2009. Internal representations, external representations and ergonomics: Toward a theoretical integration. *Theoretical Issues in Ergonomics Science*, 10 (4), 335–376.

Rouse, W. B. and Morris, N. M., 1986. On looking into the black box: Prospects and limits in the search for mental models. *Psychological Bulletin*, 100 (3), 349–365.

Salmon, P. M., Stanton, N. A., Walker, G. H., Jenkins, D. P., and Rafferty, L., 2010. Is it really better to share? Distributed situation awareness and its implications for collaborative system design. *Theoretical Issues in Ergonomics Science*, 11 (1–2), 58–83.

Santamaria, C., Garciamadruga, J. C. and Carretero, M., 1996. Beyond belief bias: Reasoning from conceptual structures by mental models manipulation. *Memory & Cognition*, 24 (2), 250–261.

Sarter, N. B., Mumaw, R. J. and Wickens, C. D., 2007. Pilots' monitoring strategies and performance on automated flight decks: An empirical study combining behavioral and eye-tracking data. *Human Factors*, 49 (3), 347–357.

Sauer, J., Schmeink, C. and Wastell, D. G., 2007. Feedback quality and environmentally friendly use of domestic central heating systems. *Ergonomics*, 50 (6), 795–813.

Sauer, J., Wastell, D. G. and Schmeink, C., 2009. Designing for the home: A comparative study of support aids for central heating systems. *Applied Ergonomics*, 40 (2), 165–174.

Sauer, J., Wiese, B. S. and Rüttinger, B., 2004. Ecological performance of electrical consumer products: The influence of automation and information-based measures. *Applied Ergonomics*, 35 (1), 37–47.

Schoell, R. and Binder, C. R., 2009. System perspectives of experts and farmers regarding the role of livelihood assets in risk perception: Results from the structured mental model. *Approach Risk Analysis*, 29 (2), 205–222.

Scholtz, J., 2002. Human–robot interactions: Creating synergistic cyber forces. In: Schultz, A. C. and Parker, L. E. (eds.), *Swarms io Inielligeni Aufomaia (Proceedings from the 2002 NRL Workshop an Multi-Robot Systems)*. Dordrecht, the Netherlands: Kluwer Academic Publishers.

Schroyens, W., Schaeken, W. and D'ydewalle, G., 1999. Error and bias in meta-propositional reasoning: A case of the mental model. *Theory Thinking & Reasoning*, 5 (1), 29–66.

Shah, N., Chen-Fang, T., Kuo-Ming, C. and Chi-Chun, L., 2010. Intelligent household energy management recommender system. In: Shiskov, B., Tsihrintzis, G. A. and Virvou, M. (eds.), *2010 Proceedings of First International Multi-Conference on Innovative Development in ICT (INNOV 2010)*, Athens, Greece.

Shigeyoshi, H., Inoue, S., Tamano, K., Aoki, S., Tsuji, H. and Ueno, T., 2011. Knowledge and transaction based domestic energy saving support system. In: Konig, A., Dengel, A., Hinkelmann, K., Kise, K., Howlett, R. J. and Jain, L. C. (eds.), *Proceedings of the 15th International Conference on Knowledge-Based and Intelligent Information and Engineering Systems, KES 2011*, Kaiserslautern, Germany.

Shipworth, M., Firth, S. K., Gentry, M. I., Wright, A. J., Shipworth, D. T. and Lomas, K. J., 2010. Central heating thermostat settings and timing: Building demographics. *Building Research and Information*, 38 (1), 50–69.

Smith-Jackson, T. L. and Wogalter, M. S., 2007. Application of a mental models approach to MSDS design. *Theoretical Issues in Ergonomics Science*, 8 (4), 303–319.

Staffon, J. D. and Lindsay, R. W., 1989. Experience with model based display for advanced diagnostics and control [of nuclear power stations]. *Proceedings of the Seventh Power Plant Dynamics, Control and Testing Symposium*, Knoxville, TN.

Stanton, N. A. and Baber, C., 2008. Modelling of human alarm handling response times: A case study of the Ladbroke Grove rail accident in the UK. *Ergonomics*, 51 (4), 423–440.

Stanton, N. A., Salmon, P. M., Walker, G. H., Baber, C. and Jenkins, D. P., 2005. *Human Factors Methods: A Practical Guide for Engineering and Design*. Aldergate, Hampshire: Ashgate Publishing Ltd.

Stanton, N. A. and Stammers, R. B., 2008. Commenting on the commentators: What would Bartlett have made of the future past? *Ergonomics*, 51 (1), 76–84.

Stanton, N. A., Young, M. and Mccaulder, B., 1997. Drive-by-wire: The case of driver workload and reclaiming control with adaptive cruise control. *Safety Science*, 27 (2–3), 149–159.

Stanton, N. A. and Young, M. S., 2000. A proposed psychological model of driving automation. *Theoretical Issues in Ergonomics Science*, 1 (4), 315–331.

Stanton, N. A. and Young, M. S., 2005. Driver behaviour with adaptive cruise control. *Ergonomics*, 48 (10), 1294–1313.

Stanton, N. A., Young, M. S., and Harvey, C., 2014. *Guide to methodology in ergonomics: Designing for human use*. Boca Raton, FL, CRC Press.

Stern, P. C. and Aronson, E. (eds.), 1984. *Energy Use: The Human Dimension*. Washington, DC: National Academic Press.

The Department of Energy and Climate Change. [Online]. London, UK: The Department of Energy & Climate Change. Available: http://www.decc.gov.uk/ [Accessed 5 May 2011].

The UK Low Carbon Transition Plan. [Online]. 2009. London, UK: The Department of Energy & Climate Change. [Accessed 10 November 2011].

Tversky, A. and Kahneman, D., 1974. Judgment under uncertainty – Heuristics and biases. *Science*, 185 (4157), 1124–1131.

Vastamäki, R., Sinkkonen, I. and Leinonen, C., 2005. A behavioural model of temperature controller usage and energy saving. *Personal and Ubiquitous Computing*, 9 (4), 250–259.

Veldhuyzen, W. and Stassen, H. G. (eds.), 1976. *The Internal Model: What Does It Mean in Human Control?* New York: Plenum Press.

Virzi, R. A., 1992. Refining the test phase of usability evaluation: How many subjects is enough? *Human Factors: The Journal of the Human Factors and Ergonomics Society*, 34 (4), 457–468.

Weyman, A., O'hara, R. and Jackson, A., 2005. Investigation into issues of passenger egress in Ladbroke Grove rail disaster. *Applied Ergonomics [Special Issue: Rail Human Factors]*, 36 (6), 739–748.

Wickens, C. D., 1984. *Engineering Psychology and Human Performance*. London, UK: Merrill.

Williges, R. C., 1987. The society's lecture 1987 the use of models in human-computer interface design. *Ergonomics*, 30 (3), 491–502.

Wilson, J. R., and Rutherford, A., 1989. Mental models: Theory and application in human factors. *Human Factors*, 31 (6), 617–634.

Xu, B., Fu, L. and Di, H., 2009. Field investigation on consumer behavior and hydraulic performance of a district heating system in Tianjin, China. *Building and Environment*, 44 (2), 249–259.

Yakushijin, R. and Jacobs, R. A., 2011. Are people successful at learning sequences of actions on a perceptual matching task? *Cognitive Science*, 35 (5), 939–962.

Zhang, T., Kaber, D. and Hsiang, S., 2010. Characterisation of mental models in a virtual reality-based multitasking scenario using measures of situation awareness. *Theoretical Issues in Ergonomics Science*, 11 (1), 99–118.

Zhang, W. and Xua, P., 2011. Do I have to learn something new? Mental models and the acceptance of replacement technologies. *Behaviour & Information Technology*, 30 (2), 201–211.

Bibliography

Brown, S. C. and Martin, J. N. T. (eds.), 1997. *Human Aspects of Man-Made Systems*. Milton Keynes, UK: Open University Press.

Denzin, N. K. and Lincoln, Y. S. (eds.), 2011. *The Sage Handbook of Qualitative Research*, 4th edn. Thousand Oaks, CA: Sage.

Edwards, E. and Lees, F. (eds.), 1974. *The Human Operator in Process Control*. London, UK: Taylor & Francis Group.

Holland, D. and Quinn, N. (eds.), 1987. *Cultural Models in Language and Thought*. Cambridge, UK: Cambridge University Press.

Norman, D. A. and Draper, S. W. (eds.), 1986. *User Centered System Design: New Perspectives on Human–Computer Interaction*. Hillsdale, NJ: Lawrence Erlbaum Associates, Inc.

Posner, M. I. (ed.) 1989. *Foundations of Cognitive Science*. Cambridge, MA: MIT Press.

Rogers, Y., Rutherford, A. and Bibby, P. A. (eds.), 1992. *Models in the Mind: Theory, Perspective and Applications*. London, UK: Academic Press, pp. 119–143.

Index

A

Action specification, 147–148, 153–154, 162, 165–169
Adaptable framework, mental models
    application of, 31–32
    bias and filtering, 15–19, 35
    content accuracy, case study of Kempton, 19–23
    definition, accuracy in, 23–31
    development of, 15–31
Alternate control devices, 99–105
Amateur theory, of home heating, 79
Analysis method, reliability
    exercise, dynamics of, 69
    improvements, 71
    inter-analyst reliability exercise, 69–71
Analysis reference table
    behaviour pattern, 55–61
    device function and behaviour, 64–65
    evaluating utility of, 65–66
    home-heating function, mental model description, 61–64
    improvements to, 65, 67
Appropriate heating control, 135
Automated data collection, 65, 117

B

Background bias, 16, 21
Background knowledge, 10, 12–15
Bainbridge approach, knowledge structures, 12
Behaviour patterns, 41, 47, 55–61, 248–255
Belief bias, 15
Bias, 39, 54–55, 82–85, 227
    background bias, 16, 21
    belief bias, 15
    cognitive artefact bias, 16, 18, 28–29, 35
    definition of, 15
    in mental models research, 49, 228
    mitigation of, 35
    risk of, 16, 31–32
    social bias, 16, 18, 21–22, 35
    subject's mental models, accessing, 20–23
    tree-ring method, 234
Bias rings, 17–18; *see also* Bias
    tree-ring profile, 22–23
Boiler activation, control of, 224–225
Boiler 'on' periods, 112, 117–119, 121

Boost button function, 148–149, 153–154, 166, 190–191
Broader system variables, 229

C

Case studies
    feedback behaviour, 95–99
    feedback mental model, 89–95
    methods, mental models
        data collection, 81–82
        dynamics of interview, 82–87
        outputs, analysis of, 87–89
        participants and setting, 80–81
    timer model, alternate control devices, 99–105
Causal association, 113
Causal model, 86
Cause and effect, 86–87, 91, 97, 102–103, 108–109
Cognitive artefact bias, 16, 18, 28–29, 35, 49
Cognitive processing, 10–11
    background bias ring, 13
    Bainbridge approach, 12
    comparison of theories of, 13–14
    Johnson-Laird linguistic approach, 11–12
    Moray theory, 12–13
Compatible mental model, 150, 152–154, 169, 202
Compatible user mental model (CUMM), 145, 186, 227, 229–231
Concept maps, 86
Conceptual models, 139–140
    of target system, 24
CUMM, *see* Compatible user mental model

D

Data categorization, 38
Data collection, 41
    from central heating system, 116–117
    development of, 49–52
    outputs, verification of, 51–52
    paper-based activities, 51
    participants observation, 53–67, 115–116
    pilot case studies, 53–55
    QuACk, 38
    setting, 116
    from user, 117

DCM, *see* Designer conceptual model
Decision ladder concept, 137
Dependent variables, 205–207
Designer conceptual model (DCM), 26–27
Design model, 140–141
    expert user mental model, 142–145
    home-heating control, appropriate, 145
    stages of appropriate activity, 145–149
Device models, 78
Domestic energy consumption, 111, 201
    reducing, 135

E

Efficient cause lattice (ECL), 13
Energy-saving, 111–134
Energy-wasting behaviour, 37
Ergonomic science, methodological issues, 38
Evaluation, activity and, 137–139, 167–169
Execution, activity and, 137–139, 162, 163–167
Expert user mental model, 142–145; *see also*
    User mental models

F

Fast-acting variable obstruction, 152
Feedback behaviour pattern, 118
Feedback mental models, 47–48, 165
    feedback behaviour without, 95–99
    valve behaviour, elements of, 89–95
Feedback model, 113
Feedback shared theories, 78–79, 87, 105, 114,
    117–118, 123, 126, 133
Final cause lattice (FiCL), 13
Folk theories, 19
Formal cause lattice (FCL), 13
Fourier's Law, 114
Functional models, appropriate, 222–223

G

Goal achievement, limitations of study,
    225–226
Goals, setting, 136–137, 145, 148–149, 153, 155,
    162–164
Gulf of evaluation, 137–139, 167–169
Gulf of execution, 137–139, 167–169

H

Holomorphic mappings, 13
Home-heating case study, core issues, 230–232
Home-heating context, assess methods, 41–47
    content analysis, 47
Home-heating control panel, domestic setting
    testing, 234–235

Home-heating interface design, 183–186, 197,
    199–200
    key devices, design of
        boost button function, 190–191
        programmer, 188–189
        thermostat, 186–188
        TRV, 191–192
    pilot run for, 197–199
    simulation, creation, 195–197
    system view, design of, 192–195
Home-heating programmer interface, 189
Human–robot interaction, 138

I

Ideographic case representations, 39
Independent variable, 205
Intelligent agents for home energy management
    group (IAHEM), 116
Intentions, stage of action, 137, 145,
    154, 162
Inter-analyst reliability exercise, 69–71
Interpretation, activity and, 149, 152–153, 167
Interview
    durations, 82
    dynamics of, 82–87
    template, QuACk, 240–246
        background experience, home heating,
            240–241
        behaviour, 241–244
        device function, mental model of,
            244–246
    transcripts, 87–89
Interviewer instructions, QuACk
    background, 237
    expectation, participant, 238–239
    interview outputs, 239
    verbal positioning, 238

J

Johnson-Laird linguistic approach, 11–12

K

Key devices design, home-heating
    boost button function, 190–191
    programmer, 188–189
    thermostat, 186–188
    TRV, 191–192
Knowledge structure, 10–11

M

Mann–Whitney test, 217
Material cause lattice (MaCL), 13
Mental cause lattice (MCL), 13

Mental model content accuracy
  case study, Kempton, 19–23
  in definition, 23–31
  definitions of concepts, comparison, 30–31
  Norman definition of, 23–26
  Wilson and Rutherford, definition of, 26–31
Mental model descriptions, 64, 256–261
Mental-model-driven design, heating goals, 230
Mental model interface design, 201–204, 221–226
  apparatus and materials, 207–208
  experimental design, 205–207
  method, 204–209
  participants, in experiment, 207
  procedure, 208–209
  results, 209–221
Meta-knowledge, 130, 154, 188
Methods, mental model interface design, 204–205
  apparatus and materials, 207–208
  experimental design, 205–207
  participants, 207
  procedure, 208–209
Method validation, QuACk
  analysis method, reliability, 69–71
    exercise, dynamics of, 69
    improvements, 71
    inter-analyst reliability exercise, 69–71
  analysis method, reliability of, 69–71
  measurement, of self-report behaviour, 67–69
Mitigation strategies, 38, 54–55, 83–85
Moray's lattice theory, 12–13, 132

N

Natural mapping, 153
Negative meta-knowledge, 130
Night set back, 79

O

Occupant behaviour, 201
Optimal consumption, home-heating controls, 229
Optimal home heat control
  complex task, 230–231
  tailored guidance for, 234
  variables, control, 232

P

Paper-based activities, 51, 82, 86
Participant information sheet, 238
Participant observation, 53, 55
  analysis reference table, 55–67
    behaviour pattern, 55–61
    device function and behaviour, 64–65
    evaluating utility of, 65–66
    home-heating function, mental model
      description, 61–64
    improvements to, 65, 67

output formats, benefits of, 64–65
  device function and behaviour, 64–65
  mental model description, 64
  self-report diagram, 64
Participant responses, 86
Payne's mental model, 31
Perception, 137, 149, 152–153, 167
Physical system lattice (PSL), 13
Potential bias, 35
Procedural semantics, 11
Processing modules, 12
Programmable thermostats, 231
Programmer devices, 153, 188–189
Programmer-scheduled time periods, 153
Propositional representations, 11

Q

Quick association check (QuACk), 37, 139–140,
    142, 145, 162, 209, 262–265
  analysis method, 52–53
  bias, in mental models research, 49
  cognitive science literature, 40–41
  content analysis, 47
  data collection method, 49–52
  development process of, 39–53
  domestic energy, exploring, 234
  energy-consuming behaviour and, 228–229
  home-heating context, assess methods for, 41–47
  human factors, literature review, 40–41
  instructions for interviewer, 237–239
    background, 237
    expectation, participant, 238–239
    interview outputs, 239
    verbal positioning, 238
  interview template, home-heating domain,
    240–246
    background experience, home heating,
      240–241
    behaviour, 241–244
    device function, mental model of, 244–246
  iterative case study approach, 73
  key elements of, 57
  method evaluation, 73–75
  method validation, 67–71
  participant information sheet, 238–240
  participant observation, 53–67
  pilot case studies, 53–55
  prototype, components of, 50
  shared theory, 38
  tree-ring method, 41, 49, 72

R

Recommendations, home-heating systems
  design with optimal consumption, 233
  mental models approach, using of, 233

system-level strategies, using of, 233
task complexity recognition, 232
Remote sensor, 132
Research, mental models concept
    aims and objectives/purpose, 2–4
    contribution to, 4
    hypothesis, 2–4
Resource-light method, 49; *see also* Quick
        association check

S

Scientist's conceptualization, 24
Self-report diagram, 64
Self-reported behaviour, 228
Seven-stage model, user activity, 137–138
Shared theories, 78, 87, 113
Social bias, 16, 18, 21–22, 35
State of system evaluation, feedback cues,
        167–168
Switch mental models, 49
Switch theories, 79
System image, of home heating, 150,
        169–180
    at device level, 152–154
    at system level, 150–152
System-level approach, 229
Systems UMM, 231–232

T

Target system, 24
Thermostat devices, 186–188
Thermostat function, 5, 37, 53, 61, 79, 87, 153,
        202
Thermostatic radiator valves (TRVs), 61, 63, 123,
        130–131, 147–149, 154, 165, 168,
        191–192, 217
    controls, appropriate behaviour, 223–224
Thermostat set point values, 117, 120–121, 123
Timer mental models, 49
Timer model, alternate control devices, 99–105
Timer theories, 79

Tree-ring method, 29, 31, 35, 49, 82, 85, 228
TRVs, *see* Thermostatic radiator valves
Tunnel vision, 193

U

UCM, *see* User conceptual model
UMMs, *see* User mental models
User conceptual model (UCM), 26–27
User mental models (UMMs), 24, 26–27, 38–39,
        78, 105–106, 108, 127–129, 139, 141,
        185–186, 188, 191, 193, 197, 200,
        204–206
    boiler activation, control of, 224–225
    compatible model, of home heating, 155
    data collection approach, 107
    discoverability, home-heating controls,
        221–222
    frost protection and holiday buttons, 223
    goal achievement, 225–226
        limitations of study, 225–226
    gulf of evaluation, 167–169
    gulf of execution, 162–167
    of home heating, case study, 154–181
    of home-heating simulation, 209–214
    self-reported behaviour, home-heating
        operation, 155–162
    seven stages of activity, home-heating system,
        162–169
    sub-hypotheses, 206–207
    TRV controls, appropriate behaviour, 223–224
    user behaviour, home-heating simulation,
        215–221
        assumption for study, 215–221
User verification, QuACk, 51–52

V

Valve mental models, 47–48
Valve model, 113
Valve shared theories, 78–79, 87, 105
Verbal positioning, 237–238